R100
IN CANADA

R100's only captain Squadron Leader Ralph Sleigh Booth in his Royal Air Force uniform with Air Force Cross and bar. Booth was awarded the bar and promotion from Flight Lieutenant for piloting the R33 home after 30 hours with the bow gas bag deflated. The ship broke away from the Pulham mast in a gale April 16, 1925 and drifted over the North Sea and the Netherlands.

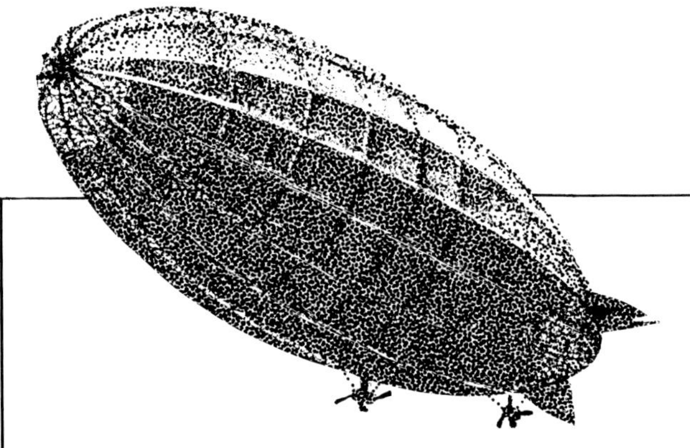

R100
IN CANADA

Barry Countryman

THE BOSTON MILLS PRESS

In Memory of My Parents

R100 in Canada
ISBN 0-919822-36-3

Published in Canada by
THE BOSTON MILLS PRESS
98 Main Street
Erin, Ontario NOB ITO

Typeset by Speed River Graphics
Printed by Ampersand, Guelph

We gratefully acknowledge the assistance of the Canada Council, the Ontario Arts Council and the Office of the Secretary of State in the publishing of this book.

Acknowledgements

I would like to thank the following individuals, corporations and archives for their help: J.H. Adams; Archives of Ontario; Peter Bailey; J. Christopher Barron; Bell Canada; Horace Boivin; Lily Booth; William O. Boss; Sir Aubrey Burke; Burmah-Castrol Canada Limited; Canadian National Railways; Canadian Pacific Railway; S.A. Carter; Chloride Metals Limited, London; City of Montreal Archives; City of Ottawa Archives; City of Toronto Archives; Arthur E. Clarke; John Robert Colombo; David Cook; A.L. Crossland; Directorate of History, Department of National Defence, Ottawa (Philip Chaplin); Georges Faille; James Fisher; Flight International; Ford Archives, Dearborn, Michigan; G.J. Garner; Roland Garrett; Victor George; Goodyear Tire & Rubber Company, Akron, Ohio; Government of Quebec Archives; the late J. Fergus Grant; Gertrude Hammond; Foster Hewitt; W.J. Holland; Jerome C. Hunsaker; The Illustrated London News Picture Library; Imperial Oil Limited; Imperial War Museum, London; Tom Jamison; John Johnston; G.W. Keep; Carol LaPierre; the late S.W. "Sailor" Lawrence; Harry Leong; Major R.K. Malott (Ret'd); Dennis Marshall; Sir Peter G. Masefield; John Mason; Emily McWhinney; the late Captain George F. Meager (Ret'd); Metropolitan Toronto Library (Mrs. N. Dainard); Sylvia Meunier; Prof. Henry Cord Meyer; Larry Milberry; Ministry of Defence, London; National Air and Space Museum, Smithsonian Institution, Washington; National Museum of Science and Technology, Ottawa; Naval Historical Center, Department of the Navy, Washington; M. Barbara North; Kent O'Grady; W.R.J. Oliver; the Ontario Science Centre, Toronto; David W.H. Pickard; Gwen Pressey; Provincial Archives of British Columbia; Public Archives of Canada, Ottawa (Carl Vincent); Public Record Office, Kew, England; The Recorder and Times Limited, Brockville, Ontario, the late Vice Admiral Charles Rosendahl (Ret'd); J.M. Ross; The Royal Canadian Geographical Society, Ottawa; The Royal Canadian Military Institute, Toronto; The Royal Canadian Regiment Museum, London, Ontario; Mrs. C.H. Rumsby; Judy Scott; the late Vice Admiral T.G.W. Settle (Ret'd); Standard Brands Food Company; Al Starkweather; E.J. Stupple; the late John Swettenham; Toronto Telegram Photograph Collection, York University Archives, Toronto; Lord Ventry; Vickers Limited, London; Vickers, Canada Inc.; the late Sir Barnes Wallis; David Watson; the late Bert S. Wemp; Reg Wilkinson; C.J. Williams; Captain T.B. Williams (Ret'd); Tom Williams.

My thanks to the Canadian Forum for permission to quote from their September, 1930 journal, to William Morrow and Company for permission to quote from *Slide Rule* by Nevil Shute and to Burmah-Castrol Canada, Imperial Oil and Standard Brands Food for permission to reproduce their R100 advertisements.

And special thanks to my patient typist George Pereira.

For copyright reasons I wish to locate the copyright holders of the New York World and the heirs of C.H.J. Snider, former managing editor of the Toronto Evening Telegram.

I have omitted the hyphen from R-100 to conform to other British rigid airship numbers.

Contents

R34, under Major G.H. Scott's command, at Mineola, New York, July 8, 1919. His later suggestion of the high mooring mast eliminated large handling parties and the gassing of an airship from hydrogen bottles.

Peacetime Airships

I have not interested myself in airships because I like their appearance, nor am I fascinated by their size, but because they are the only form of vehicle capable of traversing great ocean spaces carrying commercial loads. They are the keystone of the arch of Imperial communications.

Sir Dennistoun Burney
February 12, 1930

The British airship, the R34, piloted by Major G.H. Scott, took off on July 2, 1919 from East Fortune in Scotland and headed for New York. The flight was without serious mishap, although by the time she reached Roosevelt Field, Mineola, Long Island the gas tanks were practically empty. It took 108 hours and 12 minutes to cross the Atlantic. The R34 returned to Pulham, England July 13 in 75 hours, 3 minutes. It was the first transatlantic return trip.

The success of the R34 put Britain in the forefront of airship development but the expense of an airship development program proved a burden to the government and on March 1, 1921 the Secretary of State for Air, Winston Churchill, announced the government was discontinuing all airship activity for reasons of economy. All government airship property would be offered for sale to private companies.

The Air Ministry, as a basis for discussion with any private company interested, would hand over free its airships: the R33, R36, the only one already fitted with a passenger car, the R37, which was almost complete, the R80 and the German airships L64 and L71 which Britain had received as war reparations in 1920. Also as part of the package there was a large quantity of spare engines, fabric, gas bags and the shed and other equipment for an overseas base. The British government would also sell or lease its airship stations at Cardington and Pulham, both of which were fully equipped although Cardington had no mooring mast. Experienced airship personnel would be loaned, their salary to be paid by the company. All available technical airship data, including wireless and meteorological information and the results of experimental airship work would be placed at the disposal of the commercial company. The government, for its part, would require that British shareholders control the company's capital and that the Air Ministry be represented on the board of directors.

Despite this attractive offer the government was not satisfied with any of the commercial proposals it received. Later that year, at the Imperial Conference, Britain offered her airships as a basis for an imperial communications network, a scheme she had been interested in since 1918. Although there were several proposals put forward, notably by A.H. Ashbolt, the Agent-General for Tasmania in London, nothing came of them. The conference committee could only recommend that the best hope for the successful development of imperial air communications lay in private enterprise.

The crash of the R38 airship only a few weeks later, on August 24, further weakened British commitment to airships. The R38 had been built for the United States and was undergoing its final flight tests before being handed over to the U.S. Navy. The ship exploded over the Humber River at Hull, killing 44 of the British and American crew.

The Airship Guarantee Company

One of the most determined supporters of the commercial airship idea was Acting Commander Charles Dennistoun Burney. His proposals, backed by Vickers and Shell Oil, were submitted to the Air Council in March and April 1922. They required, as did all the other commercial airship plans, that the government turn over its airships and bases to a commercial company and guarantee financial returns for the first few years of the scheme.

For its part the new company would build a fleet of vessels each with a 5 million cubic foot gas capacity. Each airship would be approximately 760 feet long with a diameter of 110 feet. With a disposable lift of only 50% to allow for strong construction, and assuming 3,000 miles as the average length of a commercial flight without refuelling, such an airship would have 44.5 tons for mail, passengers and stores and a speed of 80 m.p.h.

The company would also construct fuelling depots and mooring bases at Port Said in Egypt, Bombay, Rangoon, Singapore and Perth, Australia. It would initiate as soon as possible a bi-weekly mail and passenger service to India with a weekly extension to Australia. The whole organization would be placed at the disposal of the Admiralty in time of war.

Burney's persistence and enthusiasm paid off in the end. On July 26, 1923 Sir Samuel Hoare (Secretary of State for Air) announced to the House of Commons that the government had accepted the Burney scheme in principle, subject to contract details being settled by the Treasury.

Before anything more could happen there was a general election on January 11, 1924. Britain's first Labour government, led by Ramsay MacDonald, took office. For the sixth time Burney's scheme was reviewed by a government committee. The government objected to the scheme's monopolistic nature, and decided that the government could probably operate the proposed airship service as cheaply as any private interest.

On May 14, 1924 the government presented its own airship program to the House of Commons. The novel feature of the proposal was the division of airship construction. There were to be two new airships built, one by the government and the other by a private company under contract. The government would also recondition one of its existing airships, the R33, for research purposes, construct an overseas terminal and an intermediate base for these two vessels to travel from England to India. The airship stations at Cardington and Pulham would also be retained.

The maintenance of two separate airship manufacturing plants and other necessary ground facilities would allow for rapid expansion of the airship program once it began. The government proposal allowed the private company, the Airship Guarantee Company, a subsidiary of Vickers, to purchase the airship at a reduced price at the end of the program. The budget for the government's scheme was £1,200,000 over 3 years and was in sharp contrast to the Burney scheme of six airships to be built over 15 years at a cost of £4,800,000.

The more important technical contract requirements for the R100, the capitalist ship, and the R101, the socialist ship, (as they were dubbed by the press) were: a gas capacity of not less than 5 million cubic feet; a speed of not less than 70 m.p.h. at an altitude of 5,000 feet; total fixed weight (the weight of the structure, the power plant and equipment) not to exceed 90 long tons; accommodation for at least 100 passengers; navigability to be possible within certain specified angles of pitch; and finally the airships must comply with the general requirements for airworthiness as specified by the Aeronautical Research Committee.

Subject to these conditions both design teams were given a free hand. Thus the R100 and the R101 were not sister ships. The price agreed upon for the completed R100 was £300,000. The Airship Guarantee Company had the right to purchase the R100 from the Air Ministry after trials for use on an approved British commercial service for £150,000.

In the fall of 1924 Barnes Wallis, designer of the Vickers airship the R80 and one of the most gifted airship designers in England, began to gather a staff together. He hired N.S. Norway, better known as the novelist Nevil Shute, as his chief calculator. Norway, in common with most of the design staff of the R100, came from the aircraft industry and had no airship design experience.

The design work began in Vickers House in London. The team later moved out to the industrial suburb of Crayford in Kent. In 1925 a small party was sent up to Howden, Yorkshire to put the property in shape for the R100's construction.

In 1921 Howden had been the last airship station to be closed. The owners had abandoned it in the face of falling prices for scrap metal. A single shed of corrugated iron, 750 feet long and 150 feet wide, stood on a steel framework rising above the marshland now strewn with the debris of wartime blimp sheds. Feathers and the remains of many hens littered one end of the shed and beneath the 7.5 acre floor the concrete trench that housed the hydrogen and water mains there was a vixen's lair.

The inside clearance of 130 feet was increased 10 feet by removing the runway at the top of the shed and a girder over the doorway. A hydrogen plant was then erected near the giant two-bay shed.

Construction of the 709 foot, 133 foot diameter R100 began during the summer of 1926 at Howden after Barnes Wallis had perfected a machine capable of making helically wound tubes from duralumin strips 11 inches wide to 60 feet long and 0.056 inches thick for the main girders. The tubes were then riveted with a helical seam. The original idea of using solid drawn tubes was quickly abandoned as 50 to 60 foot lengths could not be obtained at a reasonable price nor easily transported. The ship contained 11 miles of helical or spiral seam tubing and had a length-to-diameter ratio of 5.3:1 versus the R101's 5.5:1.

The main girders for the 2 foot, 6 inch deep, 16-sided transverse frames or rings were built up from three large tubes or booms 25 feet long, and 45 feet to 50 feet long for the longitudinal girders, joined by members of pressed duralumin to form a triangular section. Other girders were made from material 0.032 inches for the thinnest and 0.064 inches for the thickest. The transverse frames were numbered from the bow while the longitudinal girders were lettered from "A" at the top of the ship down both sides to "H"; the transverse girder between the two As was horizontal.

The frames were radially braced by wires, with an axial girder, also of triangular section, running through the inside of the gas bags from frames 1 to 13. The axial girder of square section joined the cruciform girders at frames 13, 14 and 15. Frame number 7 was the first one hoisted into position at Christmas 1926 on slings attached to the shed roof. Frames 8 and 9 followed, to be linked with longitudinal girders when several frames were in position. Work then proceeded forward to frame 1, followed by frames 9 to 15. From frame 4 to 12 all the longitudinal girders were interchangeable with regards to length and curvature.

At the Royal Airship Works, Cardington the 16 R101 transverse frames (10 feet 6 inches deep and unbraced) and longitudinals were of stainless steel; duralumin die stampings were used largely for main joints (fittings where transverse and longitudinal girders met). Between each of the 15 longitudinals there was an intermediate or reefing boom girder to support the outer cover, thus making the 30-sided R101 in section almost circular.

Unlike the R101, the R100 had approximately 50 different standard parts in the hull, planes and passenger compartment. The R101 was assembled at Cardington with the hull girders designed and built by Boulton and Paul Limited in Norwich, with other parts contracted out to two dozen firms.

As no British firm had the facilities or knowledge for making them and the Royal Airship Works could not supply both airships, the R100's 15 gas bags were made by B.G. Textilwerke GmbH of Berlin. This firm also provided double-ply bags for the Zeppelin Company, which, in turn, supplied the R100 gas valves.

The single-ply R100 cotton gas bags were lined with a double layer of goldbeater's skin (the membrane covering the caecum of cattle used to beat sheets of pure gold into gold leaf), then varnished internally and externally and sprayed with aluminum dust to protect them against the effects of humidity. The gas bags had a total area of 55,800 square yards or over 10 acres. The largest bag (no. 7) weighed 1,580 lbs. and had a capacity of 551,890 cubic feet and a gross lift of 16.7 tons.

The Airship Guarantee Company's power house and (below) silicol plant at Howden showing part of the aerodrome. The property, bought from the Air Ministry for £40,553, was 1 mile southwest of the village of Spaldington and 2½ miles north of the town of Howden.

Manufacturing the duralumin tubes in the machine shop attached to the Howden shed. Women on right are varnishing an R100 girder.

Manufacturing R100 girders in the Howden shed.

R100's fins laid out for assembly in both bays of the Howden shed.

Work proceeding on the R100 passenger compartment in the unheated shed; the roof threatened to blow off in high winds. Howden was the last airship station closed in 1921.

PAC RE-20940-4

Firemen's extension ladders allowed easy access to the 133 foot high structure. All R100 longitudinal girders from frames 4 to 12 were interchangeable.

Construction under way on the R100 passenger compartment, located between frames 5 and 6. A foreman was paid £4 a week and a worker about £2 15 shillings for a 47 hour, 6 day week. Management and the workers shared equally the value of any time or money saved. A workman's round-trip ticket from Hull to the North Howden Station on the London North-Eastern Railway cost 10 shillings for a 6 day week.

R100's bow gas bag. PAC RE-74-204

The Royal Airship Works, Cardington, England, with its two huge airship sheds (still in existence) and the mooring mast beyond. After the R101 crash October 5, 1930 the establishment began overhauling Royal Air Force trucks, work that in better times had been done elsewhere. The deteriorating economy meant Britain's new airship policy May 14, 1931 could not stop the decline in Cardington's workforce of 380, down from 860 when the R101 left for India.

PAC C-10155

Prime Minister Mackenzie King (second from left) on passenger platform of the Cardington mast, November 17, 1926, during the Imperial Conference delegates' visit to the Royal Airship Works. On his left is R101 chief designer Lieutenant-Colonel V.C. Richmond. In the 812x180x156 foot shed the bow of the R33 almost touched a replica of the R101's largest bay set up in the summer for stress tests. In the afternoon Britain's one serviceable airship disappeared in the clouds and misty weather at 600 feet — too low a height to release the two Grebe single-seater fighter planes.

Imperial Air Communications

By the time the 1926 Imperial Conference opened on October 19 Britain's imperial airship scheme was taking definite shape. The mooring masts at Cardington and Ismailia in Egypt were nearly finished and work had just begun on the Karachi shed which was to be 850 feet long, 180 feet wide and 170 feet high. Karachi, then part of India, had been chosen for climatic reasons and because it was near sea level, where the increased atmospheric pressure would allow the airship more useful lift for freight.

Imperial air communications were discussed at the conference's meeting at 10 Downing Street on October 28, 1926. Present for Canada were Prime Minister Mackenzie King, Minister of Justice Ernest Lapointe, Chief of Staff Major-General J.H. MacBrien, Director of the Naval Service Commodore Walter Hose and the recently appointed ambassador to Washington Vincent Massey.

At this meeting Sir Samuel Hoare, Secretary of State for Air, urged the dominions to commit themselves financially to the imperial airship program. He talked about the R100 and the R101 and of the considerable amount of money Britain was prepared to invest in their development. His speech proved persuasive to Mackenzie King.

The Canadian government had expressed little interest in airships before the 1926 Conference. Canada's domestic aircraft needs could not be readily filled by the airship and there was no serious lobby inside the country for an airship program.

After World War I the government had considered using blimps from Britain for forest fire patrol. It found that the airship would not be suitable for use in British Columbia because of the mountainous forest belt and that east of Hudson Bay the changeable weather could overtake the slow-moving craft. It concluded that the airship was a distinctly fair-weather craft and a costly one at that.

Transatlantic and trans-Pacific flying was of great interest to Canada. The shortest routes over both oceans from North America lay over her eastern and western coasts. The increase in trade that would occur if the airship network was successful would be very beneficial to her economy.

With trade and patriotic considerations in mind, Mackenzie King pledged Canada's support of Britain's airship program. He promised that Canada would build mooring masts somewhere in her eastern provinces and provide the necessary meteorological services for an Atlantic crossing.

PAC RE-11710-39

The St. Hubert mooring mast, September 8, 1928, still lacking the 87 ton mast head shipped from England in July. Canadian Vickers began erecting the 322.5 tons of steel July 9. Note workmen climbing the tower.

A Mast for Canada

As a result of Canada's agreement to participate in airship demonstration flights Major G. Herbert Scott, pilot of the R34 on her voyage to the United States in 1919, and A.R. Gibbs of the Directorate of Works and Buildings sailed for Canada on April 22, 1927. Their task was to report on possible sites for the Canadian airship mast.

After landing in Montreal the Air Ministry officials travelled to Ottawa on May 3 and met with J.L. Ralston, Minister of National Defence, and other department officers. The next day they visited two government properties: the Connaught Rifle Ranges, 10 miles west of Ottawa and Rockcliffe, 3 miles east of the city.

At a Canadian Club luncheon on May 7 Major Scott told his audience that the war had hindered the development of airships for commercial purposes. Four years of war propaganda had to be overcome. "When our immediate program is completed," he said, "we shall have two airships, each of 5 million cubic feet capacity, flying from England to India, and, I hope, to Canada." He stressed that imperial airship travel would be comfortable and safe. Each vessel, designed to carry 100 passengers, would have two-berth cabins, two promenade decks, a smoking room, a dining room seating 50 persons and shower baths.

That afternoon Scott and Gibbs, along with Squadron Leader A.B. Shearer, a Royal Canadian Air Force officer attached to the commission, accompanied Ralston to Halifax. Many Nova Scotians hoped that one of their towns would be the terminus of the proposed commercial airship line. The commission felt that the long train journey necessary from Halifax or Sydney westward was a serious disadvantage but no possible location could be overlooked. They drove around Halifax and visited Cape Breton Island but failed to find a suitable site. They also spent one day visiting possible sites in New Brunswick but had no more success there.

The site they were looking for had to meet certain requirements. It had to contain at least 600 acres, nearly a square mile, of well-drained, level land. The site had to accommodate the mast, the power house and gas plant and a shed, if one had to be built in the future. Other requirements were good roads connecting the station with railway centres, enough water and cheap power to operate the mooring machinery and to manufacture hydrogen and no high hills, trees or power lines surrounding the airfield.

On May 18 the commission travelled to Quebec and were met by Lieutenant Colonel J.M. Prower. Fog and pouring rain prevented them from visiting Camp Valcartier and they had to wait until the next day. The camp was the best potential site they had seen so far. It was large enough, the subsoil of sand and gravel provided good natural drainage, the area was easily accessible by road and well-served by the train and there was plenty of water nearby. Unfortunately it was surrounded by hills and the commission feared that air currents and eddies would seriously affect the mooring of airships. The area was also subject to very heavy snowfalls.

The commission proceeded to Montreal and looked at properties surrounding the city. They drove around the South Shore area between Longueuil and St. Hubert, noting several areas which seemed to be suitable for an airship station. They arrived in Toronto, the last city on their eastern Canada itinerary, on Sunday, May 22.

In an interview there Major Scott praised Charles Lindbergh's solo transatlantic flight the day before as a "wonderful stunt". For him the most notable feature of the flight was the mental strain involved in Lindbergh keeping his attention concentrated for such a length of time. But Major Scott declared that, despite Lindbergh's feat, there was no future for transoceanic airplane travel. "The test comes when a commercial load has to be carried, and it is obvious no airplane has yet been constructed capable of doing this." He also said that, unlike airplanes, airships become safer as they increase in size.

Toronto officials were very keen to have the airship base near their city and Scott and Gibbs toured many sites. The most promising was the municipally owned Jail Farm, a 900 acre tract 10 miles north of Toronto. It seemed an ideal site but despite its excellent road and rail connections, the commission decided there was inadequate level space for any future airship shed. The commission continued their search for potential sites around Toronto and returned to Ottawa on May 28.

PAC PA-34418

The Connaught Rifle Ranges in the 1920s. The federal government property 10 miles west of Ottawa would have required expensive clearing and levelling to become an airplane-airship base. Invited by the Department of National Defence, the two-man Air Ministry commission in May and June, 1927 travelled 5,800 miles by rail and 1,300 miles by road but failed to find suitable government land for the mooring mast in four provinces.

They submitted their report to J.A. Wilson, Secretary of the RCAF, on June 2. It stated that Ottawa's Connaught Rifle Ranges were preferable to Camp Valcartier if only government land was to be considered. The only possible municipally owned land was Toronto's Jail Farm, which would need a lot of work if the base was ever expanded. In considering commercial centres, suitable sites were obtainable at Ottawa, Montreal and

Toronto but the commission gave their choice as Montreal and recommended a site near St. Hubert.

Scott and Gibbs left for Detroit June 4 en route to Washington to discuss various aspects of the mooring of airships with U.S. authorities. They returned to Ottawa on June 14 for more meetings with Canadian officials. G.J. Desbarats, Deputy Minister of National Defence, wanted them to have another look at potential sites around Montreal. They wrote a supplementary report on this second visit and on June 22 Minister of National Defence J.L. Ralston called a meeting to discuss the commission's recommendations.

Ralston provisionally agreed with the recommendation of St. Hubert but suggested the commission check northwest Montreal once more. The commission obliged him but did not alter their original recommendation and it was finally accepted.

Scott and Gibbs left Montreal on June 24. They had a pleasant journey home on the Canadian Pacific liner *Montrose* despite three days of fog, and reached Liverpool on July 2.

The land chosen by the commission for the airship station was seven miles from Montreal in the parish of St. Hubert. The proposed site was a triangular shaped property containing 590 acres, all privately owned. The heavy clay soil would require underdrainage.

By September 1927 the Department of Public Works had secured options on 715 arpents (1 acre=1.2 arpents) for the St. Hubert airport at $200 per arpent. The Department of National Defence obtained 7.8 arpents from Canadian National Railways. Only one lot of one arpent was expropriated the following summer. The final cost of the land was $143,747.56.

The technical officers of the Department of National Defence recommended in early July that the British Air Ministry's suggestion that Canada place an order for the mooring tower's mast head at the same time South Africa did be accepted in the best interests of efficiency and economy. Standardization of the telescopic mooring arm and coupling cup was essential. The English manufacturers had built the only similar structures and Canada would have the benefit of the Air Ministry's experienced inspection staff during construction of the complicated machinery.

The estimated cost of the mast head, mooring arm and related machinery was £10,000. Delivery was not expected before the end of the fiscal year March 31, 1928 but funds were in the current Air Service appropriations to meet any progress payments. The tower head the Air Ministry ordered from Babcock and Wilcox Limited in Lincoln would cost Canada £6,854, less 2% because the South African head was ordered at the same time.

The R100 was not expected to visit Canada for at least another year but the airport did not stand idle. J.A. Wilson, now Controller of Civil Aviation, wrote to A.R. Gibbs outlining his plans, November 2, 1927. "We are starting a scheme of light aeroplane clubs similar to those which have been so successful in England. They will probably use the field and, in addition there are sure to be lots of visiting machines from the United States, as flying is becoming very common there and if good facilities are provided, lots of people will come up to Montreal by air from the south during the summer months."

On November 1, 1927 the field saw its first airplane as Squadron Leader J.H. Tudhope flew Wilson and Major D. Barry from Ottawa. On November 11 Tudhope arrived at St. Hubert with the first air mail from Rimouski.

In early 1928 Sir Dennistoun Burney, managing director of the Airship Guarantee Company, announced in the press that he hoped to carry out an Atlantic flight with the R100 in the fall. This timetable depended on completing the airship in the summer, carrying out trial flights within a few weeks and there being no serious modifications necessary to the ship. But if any of these conditions were not met it would probably be necessary to postpone the flight to Canada until the spring of 1929.

At St. Hubert a temporary wooden hangar had already been built, foundations dug for a permanent hangar and drainage and grading work carried out. Plans and specifications for the tower and accessories were being drawn up and contracts for this work would be let in early May when weather conditions permitted further work. The government intended to finish the mast during the summer but definitely not later than the end of September.

Despite Burney's earlier statements Major Scott considered an experimental flight to Canada in 1928 very doubtful. As he wrote, on May 7, 1928 to J.A. Wilson, the R100 might

PAC PA-48256

For some the mooring mast provided an excellent view of the aerial show marking the inauguration of international air mail service between St. Hubert Airport, Montreal, and Albany, New York, October 1, 1928.

PAC PA-119890

Some of the government officials responsible for developing St. Hubert Airport. From left: J.A. Adam, resident engineer, Department of Public Works; R.C. De Young, Acting Mechanical Superintendent, Mooring Tower, Department of National Defence; J.L. Dansereau, district engineer, DPW; A.C. Hardy, St. Hubert Airport manager; J.A. Wilson, Controller of Civil Aviation, DND; R. de B. Corriveau, Assistant Chief Engineer, DPW; and Squadron Leader J.H. Tudhope, Superintendent of Airways, DND.

Politics and R100 repairs delayed her arrival at the world's most modern mast at St. Hubert, Quebec, completed in March 1930. The mast was designed by the Department of Public Works based on the Cardington tower and Air Ministry data. The Minister of National Defence had made the department responsible for developing the site into an airport. Gasoline was pumped to the mast by water pressure because the two do not mix.

be completed in September, but it was more likely to be towards the end of the year. Owing to weather conditions on the Atlantic and the possibility of snow at Montreal, the middle of November was the latest date for the demonstration flight. Even this date was doubtful and a flight in the spring of 1929 would not take place before the beginning of May.

Work continued on the Canadian mast throughout the summer. The mooring arm and gimbal mounting of the 25.5 foot high, 25 feet in diameter mast head left London on July 3 with the balance of the structure on the 21st.

On July 27 the Department of National Defence agreed to pay F.E. Williams of Babcock and Wilcox, the manufacturers, £110 per month plus expenses for 3 months to visit Canada and supervise erection of the mast head. After consulting with Babcock and Wilcox's Montreal representative and Canadian Vickers Limited, the company erecting the St. Hubert tower and the 87 ton mooring head, Williams left England on August 11 with orders to report to Canadian Vickers in Montreal.

Arthur R. Pressey, 1944.

The Canadian authorities began to look for trained personnel to lead the mooring party. There was no Canadian military officer with any experience in mooring airships. The government then contacted the Air Ministry which suggested a Canadian officer be sent to Cardington for 3 to 4 months training as they could not spare one of its own men. Lieutenant Commander A.R. Pressey was recommended for the position and on October 1, 1928 he and three naval ratings, Chief Petty Officers Albert Parker and James Williams and Stoker Petty Officer James Temple, reported to the Royal Airship Works in Cardington for training. Chief Petty Officer George Kinch reported a few days later.

On October 11 Wing Commander E.W. Stedman, Canada's Chief Aeronautical Engineer, left Ottawa with R. de B. Corriveau, Assistant Chief Engineer of the Department of Public Works, to witness preparations for the arrival of the German airship *Graf Zeppelin* at Lakehurst Naval Air Station in New Jersey. They were particularly interested in studying the 180 foot mooring mast and machinery, the gassing facilities and crowd control procedures for possible application to St. Hubert Airport.

The 775 foot long *Graf Zeppelin* left Friedrichshafen carrying 20 passengers and 55,714 philatelic letters and postcards to help defray the cost of her flight. Stamp collectors were to play a large part in financing her future voyages.

The *Graf Zeppelin* was advertised to arrive at 8 a.m., Sunday, October 14. A large number of cars began to collect at the airport on Saturday evening. Naval ratings took charge of the parking, keeping the airport clear. The public was admitted to the airship hangar and allowed to inspect the American airship *Los Angeles* from a distance. By noon Sunday it was evident that the American officials had not planned on such a large crowd or such a delay. Food supplies were soon exhausted and thousands of people were milling about the airport, not knowing when the airship would arrive. Due to an exclusive contract with Hearst newspapers, the commander of the *Graf Zeppelin* refused to give his position to the airport officials. All they could do was wait like everyone else.

Early in the afternoon all roads leading to and from the airport were completely blocked for a distance of about 15 miles. At 5 p.m. Stedman and Corriveau decided the airship would not arrive before dark and left the airport to find some food. They were held up about two miles from the airport in a traffic jam that did not move for three hours. They reached the neighbouring town about 9 p.m. but were turned away from three restaurants before finding a hotel with any food left. They were the last party to be fed.

After their meal they tried to return to the air station to get official news but could not get past the village of Lakehurst. Police there said the authorities were clearing the station and were not expecting the *Graf Zeppelin* to arrive that night. Because of the traffic jams people parked in the bush and camped out for the night. Stedman could see fires in every direction.

On Monday morning Stedman and Corriveau went to New York, deciding to take the train to Lakehurst when definite news was known about the arrival of the airship. Just as they arrived at New York they heard that the airship had been sighted and was expected to arrive at Lakehurst at 5 p.m. As they boarded the train for the return journey the airship flew overhead and followed them the entire train journey, giving Stedman and Corriveau an opportunity to study her in detail.

Radio announcers were perched in vantage positions all around the hangar as the *Graf Zeppelin* arrived at Lakehurst at 5:38 p.m., after 112 hours in the air. The naval ratings had great difficulty in handling the crowds although there were fewer visitors than the previous day. The landing went smoothly although the airship had to wait until 2 a.m. before she could be housed in the hangar because of the wind.

Lessons learned at Lakehurst were not lost on plans to receive the R100 at St. Hubert. Stedman recommended that a network of one-way roads was needed at the airport to avoid tying up the whole district as public events such as airship arrivals attracted large crowds. The "casual buffet" selling of hot dogs and sandwiches was not sufficient, especially when there was the possibility of a long delay.

He also considered the dissemination of accurate news to the crowds a very important item. If neighbouring cities had been advised by radio that the *Graf Zeppelin* could not possibly arrive on Sunday a lot of confusion that day could have been avoided.

The British Air Council informed Canada on the last day of 1928 that it could not give a completion date for the R100. April or May 1929 seemed to be the earliest dates for the beginning of the airship's trial flights.

By the summer of 1928 the R100's transverse frames had been erected. The engines were then placed in the power cars and the control car installed. The gas bags, made in Germany of single-ply cotton, were installed in September. They were wrapped around the ship's axial girder to await inflation.

The following month saw 80% of the outer cover in place and the fins, rudders and elevators assembled on the shed floor. There was a delay in constructing the tail as failure of a member necessitated strengthening the tail assembly. Work still to be done by the end of the year included erecting the bow structure, installing the walkway, the fuel, ballast, electrical and control systems and the covering and doping of the rudders and elevators once they were in position.

Canadian officials were encountering problems with the St. Hubert tower because the Air Ministry had not sent drawings of the airship traveller or gangway for the passenger platform, thereby handicapping the design of the remote control gear for the winches. They decided to proceed with their own design and, if necessary, modify the traveller later.

Construction was proceeding normally at the airport. The radio and meteorological hut was nearing completion and an officer of the Royal Canadian Corps of Signals was installing the radio equipment. Interior work continued in the permanent hangar which was 80 feet square with lean-to additions of 20 feet. It was planned to construct a larger hangar beside it later.

There was considerable trouble with the two gasoline tanks. A small one at the temporary hangar was leaking badly and the other, of 6,000 gallon capacity, at the permanent hangar apparently had a large amount of water in it. To save much accounting and trouble J.H. Tudhope, Superintendent of Airways, recommended that all fuel supplied in future to commercial and government aircraft at St. Hubert be left in the hands of an oil company if one were willing to install equipment to the airport's requirements. The government price to commercial operators was 34¢ a gallon while the commercial companies were selling it for 30¢ a gallon. As Tudhope pointed out, "As long as this condition exists, no operator, except in the case of urgent necessity, will ever purchase gasoline from us."

Work proceeding on R100's fins and outer cover.

Shed Trials

I have no doubt as to the future of airship transport, but I have always found that mad optimism is more dangerous, and more difficult to combat, than honest pessimism.

Major G.H. Scott
May 7, 1928

For the first 5 months of 1929 the Air Ministry was still unable to give an official date for the completion of either airship.

In April the R100's remaining structural work was the erection of the bow pyramid, and mooring winch and guy rope platforms. The rudders and elevators had been covered, doped and erected. The struggle against the Howden shed to build the R100 was nearly over. In winter there was standing water on the aerodrome. On some mornings girders sheathed in ice after a heavy mist and frost meant work had to stop in the unheated shed due to the danger in climbing the ladders and girders. In all seasons the humid air in the shed, which was below sea level, frequently coated every girder in water resulting in some cases of slight corrosion. In early 1929 Barnes Wallis, the ship's chief designer, had every girder revarnished by hand to overcome the corrosion — a task which took 30 men 3 months to complete.This had set back the date of the airship's inflation. On hot days with the expansion of the roof trusses being greater than that of the ship, the girders were continually being pulled into alignment. The roof of the corrugated iron shed threatened to blow off in high winds.

Sir Dennistoun Burney told press representatives during a November 28, 1929 visit that someone had dubbed the R100 "the Cupid airship" because 20 workers had married local girls during the construction. He painted a sad picture of these homes being dismantled because there was no more work at Howden.

The R100's inflation began July 30, 1929. Care had to be taken until a bag was half full to prevent it from creeping along the axial girder which passed through most of the bags and to position the top of the bag in the wire mesh netting panels attached to the upper girders.

The initial installation was witnessed by a representative of the gas bags' German manufacturer, a Herr Strobl whom the wags called "Here's trouble"!

It was not easy to slow or stop the inflow of gas. As there was no gas holder it went directly from the silicol plant (two 30,000 cu.ft. per hour units) into the airship. To overcome the problem of the hydrogen going into the ship hot, a trench containing about 100 yards of main was flooded and a continuous supply of water run to keep the temperature of the cooling water down to about 60°F. The record day's production was 500,000 cubic feet.

Within 10 days 14 gas bags had been inflated under the direction of Barnes Wallis. Number 15 was not inflated until just before the R100 was ready for her airborne shed trials. The only setback had been a 15 foot tear in the bottom of bag 4. Someone had not slackened or removed the rope anchoring the bottom of the gas bag to the corridor as the bag was inflating.

This R100 photograph prompted one British Member of Parliament to ask, November 30, 1927, if this were not really of a schoolboy's Meccano set. Note the workmen on the transverse frame and the limited space in the Howden shed.

Tea time aboard the unfinished R100 for more than 50 MPs, businessmen and government officials, July 5, 1928. The three waitresses were fabric workers.

PAC RE-74-206

Two Rolls-Royce Condor IIIB engines of 650 h.p. each, one driving a tractor airscrew and the other a pusher airscrew, were placed back to back in R100's power cars. On the left is Harry Wilson of Goole, near Howden. The weight of the forward engine and its fairing was about 3/4 ton while the aft engine and gear box was just over one ton. Either engine could be removed at the mooring mast with a two-legged derrick and a chain purchase operated by men standing in the centre of the power car.

PAC RE-74-207

B. Nichols, foreman, watching a power car being positioned on the R100. On the right is Harry Wilson. The radiators (one shown hanging down behind the ladder) were the largest in Britain. The R101 power plants weighed over 17 tons vs less than 9 tons for the R100. But, for a range of 2,500 miles in still air the R101 required approximately 17 tons of diesel fuel at about £5 per ton vs 23 tons of gasoline at about £23 per ton for the R100.

The upper corridor leads to the promenade deck in this pre-launch photograph. The R100 had 32 cabins, each with two or four berths and curtains in place of doors. None of the cabins had hooks or shelves so a passenger had to be careful not to mix his pile of clothes with his roommate's.

R100's dining saloon with the galley on the left. The whole interior was of flammable construction. The danger in smoking was not the fear of leaking hydrogen but rather a cigarette igniting gasoline fumes in the vicinity of the engine cars. The R101's diesel engines made possible a smoking room.

Shortly, 500 soldiers grasping guy ropes and hand rails on the control car and power cars would guide the R100 along the painted centre line out of the Howden shed for her December 16, 1929 launching.

PAC RE-20940-2

During R100's engine trials water was sprayed on the floor near the power cars to eliminate static electricity the propellers generated 15 inches from the concrete floor of the Howden shed.

Due to valve testing, diffusion, losses from air temperature and air pressure variations, an airship required at least 50% more hydrogen than her maximum capacity. Prior to the R100's first flight, water, 249 tons of caustic soda and 183 tons of ferro-silicon produced 8,601,705 cu. ft. of hydrogen (20.3 tons) and 929 tons of sludge.

Precautions were now taken so that a spark from hobnailed boots or automobile exhaust did not ignite the R100's 5 million cubic feet of hydrogen. Cars were not allowed within a quarter-mile of the shed. All employees had to put on soft-soled shoes and hand over all cigarettes and matches when they clocked in at 7:30 a.m.

While the two experimental British airships were nearing completion the world's attention again turned to the *Graf Zeppelin*. Starting from Lakehurst Naval Air Station, New Jersey on August 7, 1929 under the command of Hugo Eckener, she circumnavigated the world in 12 days' flying time, returning August 29. The German commercial vessel, carrying between 16 and 22 passengers on each leg of the historic journey, landed only at Friedrichshafen, Tokyo, Los Angeles and Lakehurst.

The R100, weighing 105 tons, was now floating in the shed held down by ropes attached to earth-filled oil drums and crates of balance weights; the overhead slings had been slackened, but not cast off. With the gas bags in place work continued on the outer cover.

The engines were run whenever possible before the engine shed trials began September 25, 1929, at which time the ship's bow was buttoned into a socket in the shed door while the north doors were opened to allow the exhaust to escape. The trials consisted of running each pair of engines in the three power cars at three-quarter speed for 1½ hours, the final 5 minutes at full throttle; then the aft engine only in each car was run astern for half an hour.

The wooden propellers, 17 feet in diameter, generated static electricity 15 inches from the concrete shed floor. Water was sprayed on the floor near the power cars to eliminate this static which appeared as blue smoke from the exhaust as the ship strained at the cable drag wires. The noise of the engines within the corrugated iron shed made ear protectors necessary and communication possible only in writing.

One day, while the port engines were being run, a split in the outer cover astern of the port car lengthened from 18 inches to 80 feet, the entire distance between two transverse frames, before observers on the floor could attract the attention of those in charge of the engines. Thereafter someone was always in the control car to relay signals on the engine telegraphs from observers to stop or slow down the engines.

On November 11, 1929 the day before the Air Ministry took over the R100 for her flight trials, the lift and trim trials were carried out to determine the loads the ship would carry. One hundred men holding the handrails of the control car and power cars let go on the blast of a whistle to see if the vessel rose or sank. After adjusting the weights in the crates suspended by ropes to the ship and having the men let go several times the R100 floated motionless.

Barometric pressure and temperature readings produced specific lift (in lbs. per 1,000 cubic feet) on hydrogen gas tables. The specific lift was then multiplied by the R100's volume of gas and corrections made for impurities, mainly air, in the gas to obtain the total aerostatic lift. The fixed weights (hull structure, power plant, controls, ballast handling, ground handling and mooring gear, communication system) were then subtracted to produce the ship's disposable or useful lift for fuel, ballast, crew and passengers, food, cargo and mail.

Some consternation followed when it was discovered that the useful lift amounted to only 57 tons instead of the contract specification of 64 tons. Thus the R100 could only carry a maximum of 30 passengers rather than the 100 the contract called for.

Both the R100 and the R101, less power plant installations, weighed just over 95 tons; the original specification had called for a fixed weight of 90 tons including power plants. Originally both vessels were to be equipped with diesel engines after Howden had abandoned the development of R100 engines using a paraffin-hydrogen mixture. But, as the experimental work involved meant there would not be enough engines available when the airships were ready, the Air Ministry decided that only the R101 would have diesel engines.

Thus, the R100's demonstration flight to India was changed to Canada as it was feared that the flash point of gasoline would make a journey to the tropics dangerous.

The St. Hubert tower was nearing completion. Each of the eight tower posts rested on a concrete base 11 feet x 13 feet, 6 inches x 9 feet, 6 inches deep; with anchorage bolts and "I" beams the weight of each base was approximately 105 tons. The mast with six storeys or tiers, each about 28 feet 4 inches in height, was 69 feet 2 inches in diameter at the base and tapered to the passenger platform's 40 feet diameter at the 171 foot level. At the sixth storey level, 142 feet 6 inches above the ground, there was a platform 4 feet wide with searchlights to spot the anchorages on the ground and to facilitate night mooring. The total height of the tower was 205 feet.

The 26 foot mooring arm weighing 15 tons and operated by compressed air was designed to take a pull of 33,600 lbs. at an angle of 45° to the horizontal. It was mounted on gimbal bearings and counterweighted, through a system of cables and pulleys, to swing 30° from the vertical. When retracted and locked in the vertical position allowing an airship to swing, the arm capable of extending 7½ feet could take a horizontal pull of 44,800 lbs.

In the centre of the tower there was an elevator and stair shaft 16 feet square. (The Air Ministry intended to, but never did, enclose the Cardington mast elevator as trouble frequently occurred during high winds.) Fairleads for side cables to yaw winches were located 250 feet from the centre of the tower. There were 24 concrete blocks spaced at 15° intervals on the circumference of a circle of 750 feet radius for anchoring the snatch blocks through which the yaw cables would pass from the airship's bow to the fairleads.

At the base of the tower a two-storey gray stucco building, cruciform in shape, housed the mooring winches and heating plant. For maximum visibility to approaching airships, the mast head and the galvanized sheeting of the tower between the searchlight and passenger platforms in a diamond pattern, and the horizontal and cross bracing on alternate storeys of hollow cement blocks were painted black and chrome orange.

On Saturday, October 12, 1929 the R101, the government built airship, was handled out of her shed at 5:50 a.m. and locked to the Cardington tower. The 400-man walking-out party consisting of 200 Royal Airship Works employees, 150 Royal Air Force men from Henlow and 50 men from Bedford, had no difficulty as the wind along the ground at the shed entrance was almost nil.

The following day the ship was eased up on her wires to a height of 600 feet and remoored in about a 15 m.p.h. wind. On her third day at the mast the R101 slipped at 11:19 a.m. and proceeded on her maiden flight of 5 hours, 38 minutes around London.

Maiden Flight

On November 30, after seven flight trials of the R101, the Cardington mast stood ready to receive the R100. Gales swept the British Isles for two weeks before an Air Ministry meteorologist on duty at Howden announced on Saturday, December 14 that the anticyclonic weather with dead calm misty nights would remain settled for several days.

The next day Squadron Leader Reginald B.B. Colmore, Deputy Director of Airship Development, Major G.H. Scott, Assistant Director of Airship Development (Flying) and in charge of R100's maiden flight, Squadron Leader E.L. Johnston, the ship's navigator, Flying Officer M.H. Steff, second officer, and M.A. Giblett, in charge of the airship division of the meteorological office, arrived at the Howden shed, where Squadron Leader R.S. Booth and Captain George Meager, R100 captain and first officer, respectively, had spent all day and most of the evening preparing for the flight.

Shortly after 4 a.m. the roads were choked with trucks bringing 500 soldiers of the York and Lancaster Regiment from York who were to form the R100's walking-out party. The north shed doors were opened for the last time, the airship's Rolls-Royce engines were run slowly to warm up and the soldiers, newspapermen and her crew waited for the frosty, cloudless, moonlit night to end.

Booth gave the soldiers a short lecture on their duties; the first rule of a handling party was to maintain silence. With the first light of dawn chief engineer N.S. Norway boarded the vessel at 7:15 a.m. through the control car. Ballasting up was followed by the order at

Final work on R100's outer cover followed inflation of the last gas bag. The workforce declined from a peak of about 750 men and women at the beginning of 1929 to 300 when the airship was launched.

R100 in her Howden shed. Note the opening on the underside of the elevator fin for the legs of a man who could operate the elevator by hand in an emergency. Photograph probably taken during the November 28, 1929 press visit.

7:22 to walk the ship aft along a centre line painted on the shed floor and extending out onto the aerodrome. Plumb bobs had been suspended from the bow and stern to keep the vessel straight for the 12 parties of soldiers: four manning the control car and the three power cars; the others handled the guy ropes.

When the ship, being walked out stern first under the direction of Barnes Wallis, was about three-forths out, a slight puff of wind across the shed mouth started to move her off the centre line. But the soldiers held her steady while keeping her moving.

At 7:28 a.m. Squadron Leader Johnston announced "All clear forward". A great cheer rose from the crowd of spectators standing in the cold misty dawn as the bow of the silver leviathan cleared the shed with little room to spare.

When the ship was a safe distance from the shed Major Scott had her swung around to face southwest and checked the ballasting again before climbing into the control car. Leaning from a control car window Scott shouted to the handling party to let go and emptied half a ton of water ballast from the bow and a similar amount from the stern. Automobile horns and more cheers followed as she floated up at 7:53 a.m.

At 500 feet Booth rang "slow ahead" on two of the engine telegraphs. The ship climbed to about 1,000 feet. Her officers tried out the controls in a few slow circles over Howden before proceeding slowly with a following 6-9 m.p.h. breeze to York. The ship was performing as expected.

A course was set for the R100's new home at Cardington. Breakfast of bacon and eggs was cooked on board. There were 50 parachutes slung up in various parts of the ship for emergency use and they led to grim jokes as some passengers realized there were 57 aboard. After the first few trial flights the parachutes were removed.

With an air speed of 55-58 m.p.h. on four engines and a following breeze bringing her ground speed to about 70 m.p.h. the R100 covered the 135 miles in a little under two hours at a height of 1,700 feet. She cruised over Bedford for a while before approaching the Cardington tower. The fabric on the underside of the hull and on the lower vertical fin could be seen flapping slightly in the slipstream of the propellers.

The ship, coming in too slowly to link her wire and the mast head wire laid out on the ground, had to try again. The fourth approach was successful. The mooring cable was dropped at 12:45 p.m. and 50 minutes later the steam-operated mooring arm was retracted allowing the R100 to swing from the mast; the stabilizing rollers were attached to the stern.

R100 ends her maiden flight at her new home, the Royal Airship Works, Cardington, Monday, December 16, 1929. The Air Council allowed all accredited home and foreign news and press photography agencies to witness airship flights at Cardington.

Atlantic Flight Delays

The R100's handling and speed trials on Tuesday, December 17 — the second day of her flying career — emphasized the weakness of the pre-doped outer cover. Unlike previous airship practice of doping the outer cover when in place, the pre-doped covers on both the R100 and R101 could not be laced really taut. Pre-doping sections on special frames rather than climbing about the framework was intended to save time. But handling tended to crack the dope, thus reducing the water tightness.

The ship was near Kettering when an engine in the aft car was stopped and the control car informed that a sealing strip in the gap between the lower vertical fin and the lower rudder had come loose. Temporary repairs were completed but the ship's officers decided not to leave Bedford nor to attempt full speed trials (to convince officials at the Royal Airship Works that the R100 was 10 m.p.h. faster than their socialist airship's 70 m.p.h.) thus depriving Londoners of a view of the new airship. The R100 cruised around Bedford for three hours at various speeds.

On the speed trials of her third flight, January 16, 1930, the R100 achieved 81 m.p.h. N.S. Norway and others could not learn from the Cardington officials on board the R101's top speed. Four days later the R100 flew via London to Farnborough where a plane from the Royal Aircraft Establishment came alongside to photograph the ripples which appeared as blurred vertical bands moving along the outer cover.

Dennistoun Burney, managing director of the Airship Guarantee Company, ended Cardington's questions about the outer cover ripples, produced when the R100 flew faster than 70 m.p.h., by pointing out that the contract called for a top speed of 70 m.p.h.

The R100 carried out an endurance flight of 1,780 nautical miles on January 27-29 over England, Lundy Island and the Channel Islands. Navigation was by wireless and direction-finding stations, and fog horns of lightships and lighthouses.

The R100 was housed on January 30 after five trial flights and 87 hours, 2 minutes of flying time. Lieutenant Commander Pressey, who served as a supernumerary officer aboard the R100, was notified to report for duty at St. Hubert Airport not later than April 1 provided the date given for the R100's Atlantic flight — the second week of May — was confirmed.

St. Hubert Airport was developing into one of the world's finest airports, according to a program of matching services to the growing volume of traffic. Hotel accommodation for passengers in transit was planned and permanent roadways, runways and taxiways were constructed for use in spring and fall when the central landing field, sown to turf, was too wet for take-offs and landings.

The object of R100's fourth trial flight, January 20, 1930: an airplane from the Royal Aircraft Establishment at Farnborough photographed the ripples (which appear as blurred vertical bands) moving along the airship's outer cover. The ship's pointed tail collapsed May 22, 1930 during her sixth trial flight.

The R100 bumped her starboard elevator (in photo) against the shed door April 24, 1930 as the handling party walked her to the Cardington mast for a trial flight. The damage, which could not be repaired at the mast, delayed her voyage to Canada.

The first floor of the administration building served the airport control officers and customs and immigration department staffs, the passengers' waiting room and the post office which had originally been a hut about 8 feet by 5 feet near the mooring mast when under construction. The wireless room and the meteorological office were on the second floor.

The weather was suitable on April 24 to walk the R100 to the Cardington mast for a flight preparatory to the Atlantic crossing. But a side gust caught the ship as she was nearly out of the shed. The handling party could not prevent the starboard elevator bumping against the shed door; another five feet and the elevator would have been clear. Her officers believed that repairs could be carried out at the mast, but on closer inspection the damage was more serious than they had first thought and she was housed once again on April 27.

Wing Commander R.B.B. Colmore, Britain's Director of Airship Development, called the fin damage "very slight" in a Thursday, May 8, 1930 letter to J.A. Wilson, Controller of Civil Aviation. While it had taken several days to obtain "some special material" Colmore hoped the repairs would be completed by the end of the following week and a flight preparatory to the R100's departure for Canada undertaken in the week thereafter.

With an early morning arrival of the R100 expected on or after June 1 the Department of National Defence scheduled the official reception for 3 p.m. on the following day, preferably not a Sunday. And the city of Montreal had asked that if possible, the airship be at St. Hubert during the weekends, so that Montrealers might see her.

However there were delays and the R100 resumed her trial flights (her sixth) for engine and outer cover tests the evening of May 21 with a 23-hour flight over London, southern England and as far north as Hull. As the ship was approaching Cardington the following afternoon the officers were surprised to learn on the radio from Flight Officer George Cook, in charge of the mooring mast, that the pointed tip of the R100's stern was hanging down.

The tail fairing piece, intended to improve the flow of air at the rear of the airship, was

R100 at the Cardington mast April 26, 1930 as Germany's Graf Zeppelin *paid a 45 minute visit. Unlike the R100 with her bow mooring arrangement, the "Graf" preferred landing to the ground.*

buckled. Wind tunnel results predicting such an occurrence had been dismissed as experimental errors. Booth, in his first sole command of the R100 (Major Scott was not on board), landed as soon as heavy thunderstorms and squalls allowed. A subsequent inspection showed that the force of the slipstream had damaged the outer cover and the securing wires and tapes behind a power car.

R100 First Officer George Meager suggested to Wing Commander Colmore, Director of Airship Development, that the existing end be rounded off. This was the remedy undertaken with the ship losing 14 feet — and her esthetic appearance, according to N.S. Norway. Her new length was 695 feet 5¾ inches.

To correct the outer cover damage intermediate girders were inserted in the affected bay. But the heavier material for the fairing piece was not in stock. Calculating the time to produce heavier gauge metal, construct the new tail fairing with additional stiffening and fit it in position, the Air Council feared the R100 would not be ready to fly for another month. And because a further trial flight would be necessary the Atlantic crossing would not be until the end of June or the beginning of July. (The repairs took longer than expected and although the Air Ministry hoped they would be completed by June 21, it was undoubtedly relieved when the Canadian government requested the flight be postponed until after the July 28 general election.)

The delay in the R100's arrival did not diminish interest in the visit nor in requests from municipalities and flying clubs to the Department of National Defence for the ship to fly over towns in southern Ontario. Vancouver mayor W.H. Malkin and the city council urged Colonel J.L. Ralston, the Minister of National Defence, on June 2 to have the airship visit the West Coast.

The department wished that the R100 should spend as much time in Canada as possible and make three flights from St. Hubert. One was to Ottawa and one to Quebec City and the Eastern Townships. The last flight was a 1,250 to 1,300-mile Montreal-Detroit round trip. The route westward from Toronto would be via Hamilton, Brantford, London, Chatham and Windsor, returning over St. Thomas, Welland and St. Catharines.

Radio dealers in Montreal reported that business had picked up remarkably in the final days of July owing to interest in the R100. Almost all of the new radio sets leaving stores were sent out on approval and many out for the July 28 election campaign had not been returned.

Because so many radio stations and broadcasting chains wished to broadcast the R100 arrival from the small passenger platform of the mooring mast the government had decided in the spring to be responsible for the broadcast and deliver two separate stories — one in English, the other in French — to the Bell Telephone switchboards in Montreal where any station or chain could pick up the story. Originally the government had planned only an English broadcast but the delay in the R100's May visit allowed Bell Telephone to run a line from St. Hubert to La Presse's Montreal station CKAC, which was now responsible for the broadcast of the airship's arrival in French.

The Department of National Defence hired Thomas D'Arcy Finn, a sports announcer and the managing editor of the Ottawa Evening Citizen, and T.T. "Tommy" Shields, sports editor of the same newspaper, to do the English broadcast. For the French broadcast it hired Jacques-Narcisse Cartier, a Montreal newspaperman for La Presse and J.A. Dupont, director of radio for La Presse's CKAC. Dupont also announced the official ceremonies for the airship's welcome Saturday afternoon, August 2.

The Canadian National Railways Radio Department in Montreal, believing Canadians would not be satisfied with colourless government bulletins from the mooring mast, decided to report as much of the R100 flight as possible, following the airship up the St. Lawrence until she moored at St. Hubert.

For the 15 station coast-to-coast hook-up it hired Andy W. Ryan of the Department of Agriculture in Ottawa and Charles Findlay, a former Royal Navy wireless operator, to man a broadcasting outpost at Laval University in Quebec City; established an observation post on the roof of its downtown Montreal studios in the King's Hall Building at 1231 St. Catherine Street West; and erected a 60-foot-high wooden tower composed of two telephone poles on St. Hubert Airport property between the main entrance and the CNR excursion train station.

From the tower, W. Victor George, a programmer and announcer at CNRM, Montreal, with other duties at CNR Radio headquarters, and Captain John A. Barron would describe the R100's arrival at the mast.

The Toronto Daily Star arranged for its sports announcer Foster Hewitt to provide a third broadcast of the R100 arrival at St. Hubert for its radio station CFCA and a chain of six other stations west to Vancouver, to be carried over Canadian Pacific Railway telegraph lines.

To control the anticipated crowds the St. Hubert Airport Provisional Battalion was organized under the command of Brigadier-General W.B.M. King and Major William Baty, Jr. of the Royal Canadian Dragoons was appointed Camp Commandant. The force was made up of soldiers and RCMP.

The R100's 7th trial flight to test the outer cover and various wireless equipment began Friday, July 25 at 8 p.m. Most of the 24 hour, 805 mile tour of England, Wales and the Channel Islands was below or in heavy cloud with frequent rain. For the first time the R100 men wore a dark blue uniform of reefer pattern with gold buttons for the officers and black ones for the crew. The peaked caps with white dust covers had a badge, gilt for the officers and cloth for the men, consisting of a circle surmounted by a crown. In the centre of the circle was "R.100" surrounded by "Royal Airship Works". Red lettering was reserved for the captain.

The evening of July 27 seemed favourable for starting the voyage to Canada but the decision rested with Major Scott. After studying weather maps and meteorological data coming in from both sides of the Atlantic with M.A. Giblett, the R100 meteorologist, he delayed the departure for one day. On Monday afternoon, July 28, with R.B. Bennett's Conservatives about to win the general election over Mackenzie King's Liberals, the Department of National Defence received a cable that the R100 was leaving Cardington the next morning at 3:30 G.M.T.

Sir Dennistoun Burney, managing director of the Airship Guarantee Company, and Lady Burney said goodbye to their dinner guests Monday evening and drove north from London to Cardington on wet roads. They were accompanied by Burney's assistant, C.S. Bamber, and N.S. Norway. The weather was fine when they reached the mooring mast at about 11:45 p.m. The officers and passengers, less cigarettes and matches, were weighed in — each with a luggage allowance of 30 lbs. and 15 lbs. for each crewman — and ordered aboard the R100 at midnight.

For this the R100's 8th and longest flight her officers were Squadron Leader R.S. Booth, captain; Squadron Leader E.L. Johnston, chief navigator; Captain G.F. Meager, first officer; Flying Officer M.H. Steff, second officer; M.A. Giblett, meteorological officer; and Squadron Leader A.H. Wann, supernumerary officer (watchkeeper). Wann had been commanding the R38 when the ship broke in half over the Hull River in 1921.

From the Royal Airship Works were Wing Commander R.B.B. Colmore, Director of Airship Development; Major G.H. Scott, Assistant Director of Airship Development (Flying) and in charge of this flight; and F.M. McWade, resident inspector, Aeronautical Inspection Department.

The passengers were Sir Dennistoun Burney and N.S. Norway of the R100's constructors, the Airship Guarantee Company, who were aboard to observe the ship's behaviour and performance, and Lieutenant Commander Richard St. John Prentice of the aircraft carrier H.M.S. *Courageous*, representing the Admiralty. Prentice was replacing Lieutenant R.M. Ellis who was scheduled for the original flight in May. The flying crew would be five officers and 32 men — a normal two-watch crew. The third watch had left for Montreal by ship on May 16 and was waiting for the airship's arrival.

PAC RE-19795-9

R100 officers at Cardington with outward flight to Canada crew. Eleven men had left for Montreal May 16, 1930 under the command of R101 First Officer N.G. Atherstone.

Back Row: G.E. Long (Coxswain), L.A. Moncrieff (Cox'n), F. Williams (Rigger), H. Millward (Engineer), F. Hodnett (Asst. Steward), E.J. Stupple (Chargehand E), R. Ball (E), G. Watts (C/h E), N. Mann (C/h E), G.G. Cutts (R), T. Hobbs (Cox'n), A. Disley (Wireless), G.K. Atkins (W/T), D. Lelliott (E).

Middle Row: G.R. Scott (R), L.W. Hunt (E), A.H. Savidge (Chief Steward), W. Angus (Chief E), Captain G.F. Meager (1st Officer), Sq. Ldr. R.S. Booth (Captain), Major G.H. Scott (Assistant Director of Airship Development [Flying], Sq. Ldr. E.L. Johnston (Navigator), F/O M.H. Steff (2nd Officer), M.A. Giblett (Met. Officer), F/Sgt. T.E. Greenstreet (Chief Cox'n), S.T. Keeley (Chief W/T), J. Jowitt (E).

Front Row: A.F. Wiseman (R), F. Gaye (E), C. Flatters (R), J.F. Meegan (Chef), H. Cumley (E), C.H. Rumsby (R), J.M. Sturgeon (E), H.W. Clark (E).

A crewman's cloth cap badge. For the captain red lettering and gilt cap badges for his officers. Caps and badges were government property. The R100 civilian crew were required to pay for their jackets and trousers but apparently did not.

THE WINGED LION

—From the New York World.

An American view of the British visitor.

R100 captain, Squadron Leader R.S. Booth, wearing mittens while on elevator duty over the Atlantic on the flight to Canada.

Cardington to Montreal

With just enough light to see the field the R100 slipped from the mast at 2:48 G.M.T. on Tuesday, July 29. Aboard were 37 officers and crew, 7 passengers, and 10,440 gallons (34 tons) of fuel, giving the R100 a range of 4,500 miles, 3.3 tons of water ballast and 480 gallons of drinking and washing water. A great cheer from the crowds below told the officers that the airship was clear of the mast. Only the officers and the height and steering coxswains were in the control car. Some of the passengers had turned in immediately upon boarding and awoke at 8 a.m.

The wind was rather gusty but visibility was good and the base of what low cloud existed was at about 2,000 feet. At her flying height of 1,000 feet the 30-knot wind from the southwest meant the ship was making only 36 knots over the ground on her northwest course to Malin Head, Northern Ireland, via Liverpool.

Captain Booth was taking the first watch as watchkeeper. Normal routine prevailed as soon as the ship was under way, the riggers serving in two watches while the engineers worked in three watches.

M.A. Giblett, the R100 meteorologist, was pleased with the weather forecast as the ship headed for the English coast. Conditions over the Atlantic were considered above average for that time of year. Weather reports received on the day of departure indicated that a rather deep depression centred west of the Hebrides had remained practically stationary, while a high pressure area extended from the Azores to the Iberian peninsula. A wedge of high pressure over the mid-Atlantic in the region of Bermuda separated the Hebrides depression from shallow lows over the east coast of North America, and a low pressure trough extended from Hudson Bay to the American southwest states. This trough was likely to give thunderstorms in eastern Canada within 2 or 3 days' time.

Given these conditions the R100's course would be via Malin Head to Belle Isle or via the great circle course south of Greenland to Belle Isle, then across Labrador and along the St. Lawrence River to Montreal.

By proceeding north of the centre of the Hebrides depression some assistance from the winds could be expected once the R100 had reached the north of Ireland. It was anticipated that the northern section of the lows, now on the east coast of North America, could be used if they travelled east to meet the airship. While conditions were favourable for a fast passage to Belle Isle, with contrary light winds expected over Canada and off the coast, they did not bode so well for the final leg of the voyage.

At 3:25 G.M.T. the R100, at about 1,300 feet, was making 52 knots on four engines. N.S. Norway turned in at this time to be awoken by railway whistles as the ship, after passing close to Wellingborough, Northampton and Rugby, reached the old Roman city of Chester at 6 o'clock, just as Captain Meager relieved Booth in the control car. Norway and Second Officer Steff went out to the promenade deck in pyjamas as the ship's course was altered slightly to pass over Liverpool, now rising in a haze of smoke on the starboard. Steamer sirens and train whistles greeted the airship before she flew over the docks and the nearby red sandstone Anglican cathedral at 6.20. Two minutes later she crossed the English coast at Formby Point headed for the Isle of Man in the Irish Sea on a northwesterly course.

Norway went back to bed but was called at 8 a.m. to pump gasoline just as the airship emerged from a rain cloud. Pumping fuel was a half-hour chore for two men every two hours. Hand-operated Zwicky pumps transferred fuel and oil from storage tanks to gravity-feed tanks above each engine car.

Spotting the two-peaked Sugar Loaf Rock near the south end of the Isle of Man the officers altered course to run parallel to the east coast and pass by Douglas, the capital. A few minutes past 8 the ship was over the Point of Ayre at the north end of the island. At 8:40 she passed the Mull of Galloway at the southwest corner of Scotland. It was cloudy at about 2,000 feet with occasional rain but good visibility generally and a calm sea.

She was almost following the route of Postal Packet 162 which left her 300 foot high tower in London one winter evening in the year 2000, crossed the English coast at Bristol, traversed the Atlantic, and followed the St. Lawrence River to dock at Quebec City in Rudyard Kipling's 1905 short story "With the Night Mail."

The R100 maintained a course to the northwest as the wind, still from the west, indicated the airship had not yet passed to the north side of the depression. Trawlers hooted greetings as she headed for Inishtrahull, an island off Northern Ireland, while still cruising at 1,500 feet on four engines and making about 50 knots. At 9:30 the R100 passed the Glasgow-Belfast mail boat and within a few minutes sighted a lighthouse and a wreck near the group of rocks known as The Maidens.

About this time Squadron Leader Booth donned surgical rubber gloves and began exposing Petri dishes outside the control car window. During each watch of the outward flight the officer on duty would set out a dish. The Cambridge School of Agriculture wished to ascertain whether vegetable organisms or minute living spores existed in the upper air over the Atlantic. Upon the R100's arrival in Montreal the small flat glass discs were sent back to Cambridge to see if microscopic organisms survived on the gum agar-agar preparation at the bottom of the dishes.

So far there had been no mechanical problems, but the outer cover of the airship was wet from the rain clouds and fog. Temperatures were mild. A view of the north Irish coast at 10 o'clock was cut short 15 minutes later by a bank of low white cloud. Five minutes later when it began to rain the ship was abeam of Rathlin Island in the North Channel which separates Northern Ireland and Scotland. At 10:30 she was passing the centre of the depression.

Sights for a position line were taken at the R100's point of departure off Islay Island, Scotland at 11 o'clock. A 15-knot following wind indicated the airship was now north of the depression; the cloud broke up and the wind became more northerly. The weather was sunny with a clear shadow of the airship on the blue sea. The vessel's speed was reduced to 45 knots on the forward engines and the aft engine of the centre line car was stopped. She had left Cardington with 34 tons of fuel, which was good for 92 hours at her present rate of consumption. At noon the ship's clock was put back one hour.

The great adventure had begun with everyone full of confidence. But a small cocktail party in the dining saloon to celebrate leaving Europe behind was premature. Low cloud, the top of which was at about 1,000 feet, hid the island of Inishtrahull and Tory Island lay further on the ship's course. At 11:25 zone time Tory Island was about 5 miles off the port side.

The meteorological chart made at 12 o'clock was poor, showing beam winds for the next few hundred miles and probably head winds after that. With the wind at 030 degrees true 18 m.p.h. a course was set west, true, to bring the northerly wind on the starboard quarter to set the ship to the southwest.

Lunch consisted of tomato soup, stewed beef, peas, potatoes, custard, beer, cheese and coffee. The ship was carrying 1,918 lbs. of ordinary rations for three days for 48 persons, two days' ordinary rations carried in reserve, and one day emergency rations (88.5 lbs.) consisting of preserved meat, biscuits, tea, sugar, chocolate and chewing gum.

Airship Communications

R100 communication was divided into four stages. During each stage there were certain routine times both for the transmission and reception of messages. The radio station at Cardington (call sign GEC) would be in touch with the airship for the first few hundred miles and then hand over communications to Rugby (GBV) where the powerful Post Office transmitter would maintain contact for the first 1,000 miles of the voyage or until she reached 35°W longitude.

Any communication from the R100 addressed to Canada would be sent to Cardington and retransmitted over the Marconi shortwave Beam service. West of this longitude the airship would be in direct communication with Canadian Marconi's longwave transatlantic station at Louisburg, Nova Scotia (VAS), as St. Hubert (VFN) was not considered sufficiently powerful to handle the traffic between mid-Atlantic and Canada.

All air-to-ground communication would be carried over Canadian National Telegraph lines from Louisburg to the Marconi offices in Montreal. Communications from the airship coming through Louisburg but destined for Cardington would be sent from the Marconi office over the beam system. About the time the R100 reached the Canadian coast the St. Hubert radio station expected to take over all traffic with the ship.

When within approximately 150 or 200 miles of St. Hubert, as the captain decided, communication with the base would change from telegraph to radiotelephone and control placed in the hands of the landing officer on the passenger platform of the mooring mast.

As it was doubtful that existing wireless facilities could reliably transmit meteorological information between St. Hubert and Cardington the Canadian government wished that the demonstration flight utilize existing commercial cable and beam wireless services.

The Marconi Beam service, inaugurated October 25, 1926, provided direct communication between St. Hubert and Cardington. Canadian Marconi's transmitting station was at Drummondville, Quebec, 30 miles east of Montreal, and the receiving station at Bridgwater, England. The company's receiving station at Yamachiche, 25 miles north of Drummondville, received reports of the R100's progress from the transmitting station at Bodmin, Cornwall.

Although the beam system could transmit up to 200 words per minute commercial cable (Canadian National Telegraphs and Canadian Pacific Telegraphs), which seldom exceeded 50 words per minute, was also employed for base-to-base traffic as atmospheric conditions could sometimes affect radio channels. A teletype line was installed by Bell Telephone between St. Hubert Airport and the Marconi offices in Montreal; similar arrangements linked Cardington and Bridgwater, Somerset.

The R100 had two wireless transmitters. The T.X.15, the main transmitter, was adapted for continuous wave telegraph transmission only. The set, with a range of 600-4500 metres and an output of approximately 500 watts, was used almost continuously on the outward and homeward flights.

The T.22 (500-3000 metres) was an auxiliary set used for telegraph transmission and for voice transmission when in the vicinity of either the St. Hubert or the Cardington mast. This set was driven from the ship's batteries and was therefore a standby transmitter in case power was unavailable from the mains. The T.22 was used exclusively during the first 6 hours of the R100 voyage and frequently for the remainder of the outward flight for taking bearings from merchant ships.

The airship also had two receivers. The R.X.18a (300-25,000 metres) could be used for both telegraph and telephone reception. Unfortunately, the long- and medium-wave receiver was unselective on 1176 metres, requiring many requests for repetitions from Cardington. Difficulty in receiving Air Ministry traffic on 4098 metres made meteorological transmissions unreliable.

The R.X.36, a shortwave receiver provided in case of an emergency, was satisfactory on all waves in its 20-100 metre range but could not bring in 18 metres for meteorological reception. A coil made up by the Royal Canadian Corps of Signals in Ottawa for the R100's return flight for 18 metre work failed to oscillate on 18 metres.

The R100 wireless equipment was mounted in two cabinets, both provided with enclosed flash-proof and ventilated compartments. The weight of the complete installation, including spares, was 1,020 lbs.

Meteorological telegrams between the two bases were prefixed with the word Meteor, i.e. Meteor, Airships Bedford or Meteor, Hubert Montreal. Because of her heavy wireless schedule the R100 (call sign and registration number G-FAAV) wavelengths were not disclosed in the hope that private stations would not attempt to contact her.

Wing Commander R.B.B. Colmore, the Director of Airship Development, was responsible for the R100 progress reports dispatched to Cardington from the airship every 6 hours. They contained the position of the airship at the time of the dispatch, her speed since the previous report, the proposed course on the next section of the flight, the type of weather encountered, and any news of interest. Each message was prefixed with a code word which indicated the number of engines at cruising speed and if any were temporarily or permanently out of order.

The reports were released to the press as soon as they were received except for those parts clearly indicated official information. Colmore also kept a press log which was intended to be a narrative of the flight.

By the middle of the afternoon the sea was calmer. There was no sign of life. All the birds had been left behind and no ship had been sighted since leaving the Scottish coast. All on board were settling down to ship routine. Card playing and sleeping were the most popular methods of passing the time. Most of the passengers and off-duty officers had an afternoon nap; the engine noise was about the same as that of a ventilator fan in a steamship. Everyone was sucking candy in lieu of smoking.

It was not necessary to change into warm flying suits nor switch on the electric radiators in the passenger compartment.

Around 5 o'clock a whale was sighted on the port side. Now 16 hours into the flight at 54°N, 15°W the R100 was making good about 44 knots. A rough calculation indicated she should reach Newfoundland by the evening of the next day.

At 53°30'N, 23°10'W, 960 statute miles from Cardington, the static was so bad that the R100 could not read the base's weak signals. The Rugby schedule was then adopted. The R100 would transmit on 1176 metres to Cardington and listen in on the Rugby wavelength of 3846 metres from 40 to 55 minutes past each hour and for 10 minutes at 25 minutes past specific hours for special meteorological reports from Atlantic shipping compiled by Cardington. The first signal from Rugby at 3:55 G.M.T was Q.R.U., "I have nothing to communicate."

Once the R100 lost contact with Cardington at 2:35 G.M.T. she was out of touch with both sides of the Atlantic for 9 hours and 650 statute miles until she picked up Louisburg at 11:48 G.M.T. From information supplied by ships in the area she located the low that had been forming over Bermuda for the previous two days; it began to move eastward and hit the western circumference of the high. Early Wednesday morning the wind, after decreasing, began to veer and increase, eventually reaching 32 m.p.h. from the southeast. At midnight the clock had been put back one hour.

At daylight the R100 was over a dense layer of stratus clouds at 1,200 feet. Soon afterwards she ran into a clear patch and at 5:10 a.m. zone time sighted Cunard's *Ausonia* about 5 miles away bound for Montreal — her first sighting of anything since leaving Islay Island, Scotland. There was great excitement among the passengers and crew of both vessels. Wireless greetings were exchanged and signal flags hoisted on the steamer which had sailed from Southampton the previous Friday. Judging by the white crest of steam from her whistle the captain was determined to awaken those passengers who had not yet appeared on deck to greet the airship.

On four engines she was making considerable northerly progress in order to get a following wind by going round the top of the depression and reached a ground speed of 73 knots for several hours. After breakfast the forward engines of the port and starboard power cars which had run continuously were stopped and examined. The forward and aft engines in the bottom car were started up to replace them. Some spark plugs and a rocker-arm bushing were changed but otherwise the reconditioned Royal Air Force engines were in excellent condition.

The ship was running in thick fog and tried unsuccessfully to get under it by descending to about 700 feet. The fog, which lay at 500 feet, made the ship very wet. Ridges of stiff fabric stuck to the outer cover near the upper fin collected 1.5 tons of water, giving the vessel 7.2 tons of water in the ballast bags in the walkway at the bottom of the keel. She could carry 10 tons of emergency water ballast.

At 8:48 a.m. in 54°15'N, 38°30'W the R100 established contact with Louisburg, Nova Scotia 1,120 statute miles away, transmitting on 1760 metres and receiving on 2800 metres from 40 to 55 minutes past each hour.

The R100 was proceeding at 1,200 to 1,500 feet through fog, occasionally breaking through to clear sky. When the sea was visible a Hughes periscopic drift sight and a paper cube of aluminum dust determined wind direction and velocity and the ship's ground speed. The periscopic fitting enabled the observer to keep his head out of the slipstream.

The sun through the fog and clouds made such a bright light that a curtain was rigged to protect the steering coxswain's eyes from the glare, while the officers in the control car wore dark glasses.

Arthur Eldridge, Colmore's official secretary acting as the captain's clerk, organized and won the sweepstake on the day's run, noon to noon G.M.T. with 1,095 nautical miles.

In the middle of the afternoon clouds extending from sea level to about 1,500 feet prevented the R100 from seeing any ships in the area. She passed into another time zone and put the clock back one hour — after tea. The cakes served at tea time were baked on board. Suppers were cold so as not to keep the kitchen staff up too late. Before turning in, passengers had a beer, rum or a whiskey and soda; the barrel of beer ran out Thursday morning.

The R100 was facing an anticipated 20 knot wind and running on all six engines, the forward ones at 1,500 r.p.m. and the aft ones at 1,600 r.p.m. producing about 58 knots. Although there was a loss of propeller efficiency in having a pusher airscrew in the slipstream of the tractor airscrew, there was a gain in weight in having two engines in one power car and also a reduction in the number of the engine car crew.

As the R100 approached Belle Isle navigator Squadron Leader Johnston announced that a light would appear about 5 degrees from the island. Five minutes before they should have sighted it the officers looked out at 8:45 p.m.; the Cape Bauld lighthouse was ahead on the port bow. Without altering course the ship steered for Belle Isle.

The R100 received detailed weather reports every 6 hours from St. Hubert. John Patterson, Director of the Dominion Meteorological Service of Canada in Toronto, detailed Frank O'Donnell, his chief forecaster, to head the 7-man St. Hubert office for the R100 visit. The Toronto office compiled weather reports (barometric pressure, air temperature, wind velocity, humidity and cloud appearance) from 13 eastern Canada auxiliary balloon stations — established after Canada agreed to construct a mooring mast — and ships on the western side of the Atlantic and teletyped the information to the special operators at St. Hubert maintaining a 24-hour watch.

A teletype line installed for the air mail service linked Rimouski, St. Hubert, Kingston, Toronto, Hamilton, London and Windsor to transmit routine weather reports from each station to all other stations every day. St. Hubert copied the regular radio weather broadcasts from Albany, New York and the U.S. Weather Bureau via the Navy radio station at Arlington, Virginia and made up a daily weather map of the eastern half of North America.

Patterson was no stranger to the use of meteorology to meet the needs of aviation. Ten years before he had visited all Canadian Air Board stations to give instruction in the use of pilot balloons for measuring upper air currents. Daily observations were kept and forwarded to Toronto. He had inspected the R101 at Cardington August 22, 1929 while in London for a conference of empire meteorologists.

At 9:15 p.m., on Wednesday, July 30 the R100 was over Belle Isle, just being able to see it through a hole in the fog which shrouded the strait up to 1,500 feet. It was a pitch black night and visibility was nil. To avoid any high land and because the coasts were badly mapped the airship kept well in the middle of the strait.

The shore-to-shore crossing from Ireland to Newfoundland was 37 hours, 39 minutes. It was 46 hours, 27 minutes since she had left Cardington. The R100 became the fourth dirigible to make the east-to-west crossing of the Atlantic — following the R34 in 1919, Germany's LZ126, which became the American *Los Angeles*, in 1924 and the *Graf Zeppelin* in 1928 and 1929.

St. Hubert had expected to take over communications from Louisburg when the airship had reached Belle Isle, but this was not possible due to thunderstorms in the Gulf of St. Lawrence.

With plenty of fuel the airship ran down the Strait of Belle Isle on six engines on a tail bearing from the Belle Isle Direction Finding station. By dawn Thursday, July 31 she was in the Mingan Passage between the north coast of Anticosti Island and the Quebec shore, still cruising on six engines and making good 40 knots. At 6:25 a.m. the R100 began calling St. Hubert. While signals were exchanged it was not until 8 o'clock, when the R100 was at the western end of Anticosti Island in fine weather and a blue sky, that St. Hubert, 520 miles away, could definitely take over control from Louisburg. Enthusiasm aboard was running high; the sweepstake on the day's run gave way to bets on the time of arrival.

Shortly after sunrise at 4:41 a.m. the same day the quiet of St. Hubert Airport was broken by the arrival of 70 Montreal policemen on motorcycles. Bugles sounded to awaken the soldiers. Later in the morning trains brought the first of the crowds from the metropolis to join those trickling up the roads to the main entrance.

A tri-motored cabin plane left Rockcliffe Aerodrome, Ottawa at 10:50 a.m. bringing G.J. Desbarats, Deputy Minister of the Department of National Defence; Major-General A.G.L. McNaughton, Chief of the General Staff; Group Captain J.L. Gordon, Director of Civil Government Air Operations; Group Captain E.W. Stedman, Chief Aeronautical Engineer; Flight Lieutenant Reg Grant of the Aircraft Inspection Department; and R.H. Hadow, Britain's acting High Commissioner to Canada.

The officials, arriving at St. Hubert at 11:55 expected the R100 around 4 p.m., with a landing after 6 o'clock if the sun was strong. J.A. Wilson also arrived by air that morning. Ralston, the Minister of National Defence, arrived by plane from Boston before 1 o'clock.

In the northern section of the airport the final test of the airship mooring equipment was taking place. Lieutenant Commander A.R. Pressey was directing the operation from the control panel on the passenger platform of the mast. Beside him stood Lieutenant Commander Charles Rosendahl, the former captain of the American airship *Los Angeles*. Far out in the field stretched the main and two guy cables as the mooring crew stationed at proper points gave them an idea of how the $376,000 high mast could be expected to work. The hiss of compressed air as the mooring arm projected and withdrew, the clank of winches and the ringing of signal bells proclaimed a successful test.

The St. Hubert main hangar had been converted into a press gallery. Here approximately 200 newspapermen, and one woman from a New York daily, pounded out copy for all parts of the world. The gallery was served by over 40 telephone and wireless operators. All Canadian telegraph and cable companies, as well as the Canadian Marconi Company, had established offices there.

The airport's public address system was installed by the Northern Electric Company. Groups of horns were placed at the fire station, on top and inside the large hangar which served as the press gallery and near the junction of De la Savane Road and the road leading to the mooring mast.

Arrangements were made so that either the English or the French story of the R100's progress could go out over the public address system. La Presse's CKAC provided music from their Montreal studios.

Many St. Lambert, Longueuil and Montreal families, hoping to catch a glimpse of the R100, parked along the roadsides for picnics. Small boys and girls on summer holidays were among the first arrivals at St. Hubert. Their bicycles lined the airport fence, for miles it seemed.

More than three times the average number of planes were registered at the airport. Most came from the New England states with practically every air field in the northern United States represented.

At 12 o'clock the R100 passed into her last time zone; her clock was put back an hour to 11 a.m. Eastern Standard Time. Judging from the smoke of the steamers that the head wind was less strong nearer the south shore the ship edged in, at once increasing her ground speed by 6 knots to about 45 knots.

At 12:45 p.m. (Eastern Daylight Saving Time) the R100 was 2 miles off Father Point. A Loening amphibian which had kept her company for a while landed at the nearby Rimouski aerodrome; several airplanes took its place.

Two of the largest gas bags had shown signs of leaking during the voyage. Climbing along the radial wires between the bags a party of riggers under coxswain T. Hobbs found and mended three 3-inch slits along the central radial wire in bag 7 and two in bag 8. To reach the holes the airship rose to 3,000 feet to bring the bags to the riggers. From that height binoculars revealed foam marks to be three porpoises playing together.

Storm Delays

The officers wanted to land before dark but at 3:40 p.m., when the airship was just past the mouth of the Saguenay River she was struck by a sudden white squall from high hills on the north shore. In the two or three heavy bumps she oscillated rapidly over about 10° in the worst motion experienced in her 7 month flying career. The ship had increased her cruising speed from 58 to 60 knots a minute or two before the squall struck. She now headed to the south shore of the St. Lawrence and was soon out of the disturbed air but reduced her speed anyway.

At the first bump Captain Meager, the first officer, was washing his hands. He immediately went down to the walkway and met N.S. Norway just before the passenger compartment. For a few seconds they watched for any movement of the sling wires which held the three-deck compartment to the hull between frames 5 and 6. Not detecting any movement of the cables Meager proceeded to the control car to learn of any damage reports.

With no more bumps the ship's speed was increased momentarily to 58 knots. The starboard and aft car telegraphs rang for assistance. Meager, asked by Major Scott to investigate the calls for assistance, hurried aft with Norway and Squadron Leader Wann. In the lower fin they spotted two tears about 3 feet long where two tapes had torn away. Meager detailed assistant coxswain L. Moncrieff and rigger A.F. Wiseman to repair the fabric which, if left, would be whipped into ribbons by the airstream.

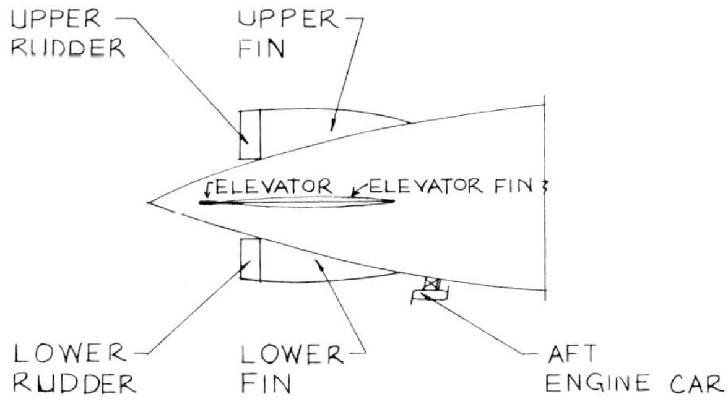

Meager and Norway proceeded up a transverse girder to the starboard fin. Here the outboard edge of the 125 foot long fin was split longitudinally for 10 or 12 feet on the lower surface. Meager went down to get help while Norway climbed out along 14 finpost (a cruciform girder) into the backbone and managed to pull together the beating fabric and stop it from splitting further until he was relieved by Meager and coxswain Hobbs and rigger F. Williams armed with fabric, needles, thread and dope. The ship reported her trouble to St. Hubert and said it would probably preclude a landing that night.

Although no damage report had been received from the port wing car Meager decided to check both the port and upper fins. He asked Norway to look at the port fin while he climbed up the cruciform girder to inspect the upper fin. With no daylight showing through the fabric Meager made his way along the horizontal arm of the cruciform girder to the port fin and discovered what Norway had found at frame 15, a hole large enough to drive a double-decker bus through. It extended on the lower surface from the radius to the backbone and was about 15 feet by 12 feet. The tapes were sound but the fabric just aft of the leading edge was in ribbons. Meager hurried down to the walkway and reported to Squadron Leader Booth in the control car from a nearby speaking-tube. Although Norway thought the hole could not be repaired Meager suggested to Booth running lines of cord across the opening and stretching canvas over the hole. He asked Booth to send down chief coxswain T.E. Greenstreet and as many riggers as he could spare.

Greenstreet appeared on the walkway a few minutes later carrying a large roll of cotton fabric. Disregarding orders to take out the corridor covering to lighten the airship for the voyage, he had stowed it in his locker thinking it might come in handy. After sewing and sealing the starboard and lower fins the ship's 8 riggers, except assistant coxswain G.E. Long who was on duty in the control car, began to repair the port fin, some standing on wires to which the fin fabric was laced.

After two hours the ship was able to proceed at 40 knots with little flapping of the cotton sheet in the port fin. A crew under Moncrieff and Long with flashlights and sometimes working face down, continued repairs until midnight on the fin covers which were almost rotten in places and quite porous.

Hundreds of Montrealers were hastening to St. Hubert. By 5 o'clock De la Savane Road bordering the airport was crowded with cars and pedestrians. Throughout the afternoon throngs of spectators, aware that they could not enter the airport until the R100 was moored, continued to surge about all entrances. At 5:10 the ship was over Grosse Isle, 30 miles below Quebec City.

At 5:42 p.m. came the opening announcement by Victor George from CNR Radio's observation tower at St. Hubert. The airship radioed St. Hubert at 5:55 that temporary repairs had been made and she was making 20 knots against a strong head wind. Lieutenant Commander Pressey replied that with the airport's floodlights the R100 could make a night landing.

The government broadcast a four-minute summary of R100 activity in English and French at 6 o'clock. Victor George and Captain John Barron spoke for 16 minutes, followed by the music of Paul Whiteman and his band and a one-minute announcement at 6:28 p.m. about the R100's fin damage.

At 2 o'clock the broadcast by the Toronto Daily Star's Foster Hewitt on behalf of 7 radio stations started and stopped as the R100 crew was repairing the damaged fabric. The weary wait continued for the landing party, the government reception committee, invited American guests, radio listeners and many spectators milling outside the airport entrances. In hangar 3 the press contingent, armed with binoculars and pocket cameras, passed the afternoon by playing bridge; the no-smoking ban was enforced by the scarlet-coated Royal Canadian Mounted Police.

Hewitt intended to broadcast the mooring operation from the press headquarters as no reporters were allowed on the field. Because the hangar was too noisy he phoned his commentary to Charles Jennings, a freelance broadcaster, at the CP Telegraphs office in Montreal. It was Jennings' voice which went over the wires during the afternoon and evening.

Doubting that the R100 would land by 8 p.m. and to learn more about the fabric damage announced rather cryptically at St. Hubert around 5 o'clock, J. Fergus Grant, aviation and marine editor of the Gazette (Montreal), got Captain A.F. Ingram, manager of Canadian Airways Limited, to fly him down to Quebec City to meet the airship. At 6:06 p.m. off the eastern end of the Island of Orleans they saw the British visitor.

She was making about 40 m.p.h. with two of the six engines stopped, but a head wind reduced her ground speed to about 20 m.p.h. The 15 foot tear in the port fin was not noticed until the mail plane had flown along the port and starboard sides and several times around the stern.

Stopped for fabric repairs over Riviere du Loup, July 31, 1930, 50 miles from Quebec City. The Riviere du Loup River on the south shore empties into the St. Lawrence River, at bottom.

The plane had kept its distance for fear of causing concern to the R100 captain but did fly within less than half a mile when inspecting the damage, thus violating the government directive of not approaching within two miles of the vessel. After 10 minutes Grant and Ingram began their return to St. Hubert leaving behind two or three picture planes. The British leviathan, her bow still rising and falling in the bright late afternoon sunshine, was 160 miles and many hours from the same destination.

Andy Ryan began his CNR Radio report from Quebec City at 6:40 p.m. as the R100, 10 miles away, loomed larger and larger on the eastern horizon. Hundreds had thronged various vantage spots from early morning to watch the R100 pass over the city. Her arrival later than anyone expected did not affect the size of the welcoming crowds.

Learning that the airship was due over the town in 90 minutes Ryan and Charles Findlay scurried around to find Laval University at 7:30 a.m. Once there they located the telegraph operator who had been sent in advance to set up the broadcasting equipment and found he had placed it and the microphone in the basement of a university building.

Ryan discovered that by hanging out of a first-floor window he would have only a 10-second view of the airship. The three men moved all the equipment to a glass-sided tower atop the 6-storey building providing an excellent view of the university and the R100's approach.

With his shortwave receiver Findlay picked up the Marconi Beam message to England indicating the R100 was in difficulty below Quebec City, then later spotted the airship through his binoculars as she turned eastward again behind the Island of Orleans.

Two hours later the boat whistles and fog horns from the St. Lawrence River joined hundreds of automobile horns, factory whistles and fire sirens in Upper Town as 40,000 people crowded Battlefields Park and lined the ramparts of the Citadel, Dufferin Terrace and the narrow streets of Lower Town cheering and waving at the silver leviathan. The ship, making only 25 knots in the face of head winds, gracefully dipped her bow.

Ryan's microphone picked up clearly the sound of the R100 engines as they drowned

Making only 25 knots against head winds, the R100 is approaching Quebec City, July 31, 1930, the first large Canadian town to see her. On the left is Canadian Pacific Railway's Chateau Frontenac Hotel. Across the St. Lawrence River (on the right) is the town of Lévis.

out the roar of the crowds. The broadcast excited all listeners and continued until 7:05 when, moving slowly in the light of the setting sun over the St. Lawrence, the R100 passed out of sight over the Quebec Bridge several miles west of the town. To N.S. Norway the town of 130,000 appeared smaller than he had expected, but was happy that the ship's relatively sound starboard fin was towards the Quebec capital!

As darkness fell at St. Hubert storm clouds gathered in the west. Some automobiles began to creep back to Montreal and nearby towns and villages. Reports from pilots flying in to St. Hubert from points in the Ottawa Valley showed that a thunderstorm was approaching and would likely cross the St. Lawrence River near Trois-Rivières, halfway between Quebec City and Montreal.

The R100 was advised at once by radiotelephone to turn south at Trois-Rivières to avoid the storm, which to her officers was a bank of clouds slightly bronze in colour and raining underneath. Before going up to the dining saloon with Norway, Major Scott ordered the ship to go through the storm rather than around it.

At 8:40 p.m., just before the town, the airship, making 40 knots, went between two heavy clouds. Burney, Norway, Booth and Scott were drinking sherry in the dining saloon when the first pitch of the approaching squall was felt. The gust under the bow put the ship's nose up and as the squall moved aft it got under the stern and tilted the ship.

Booth and Scott went down to the control car while Burney and Norway headed for the promenade deck. The vessel then hit a violent bump. Elevators were put down hard to check the rise until she reached an angle of about 20° bow down. In that position she was carried from 1,200 feet to 3,000 feet in about one minute and hidden in the clouds while swinging 8 points from her course. The ship steadied briefly before a more violent gust, which Norway calculated had a vertical speed of 50 m.p.h., shot her up another 1,000 feet in 15 seconds, with the bow down at an angle of about 25°.

A crewman in the crew's quarters above the wireless cabin had knocked over a 5-gallon drum of red abide dope, used for the fin repairs, then slid on his seat along the deck in it.

Following the St. Lawrence River to Montreal, R100 over the Quebec Bridge several miles west of Quebec City, 7:05 p.m., July 31, 1930.

When the flammable dope began leaking into the control car Major Scott ordered the lights turned off to prevent the risk of fire. Thus, with only the instrument panel bathed in dim orange light the officers were unable to note the rate of rise by the climb-meter or how long the bumps lasted; the liquid in the pitch indicator had disappeared beyond the 20° limit either way.

Food and pots and pans in the galley were upset; the dinner of soup, meat, vegetables and cakes laid on the centre table of the dining saloon shot off down the staircase and up the walkway until some of the food and cutlery, found the next day, reached frame 2.

In the 10 minutes of darkness it rained so hard that almost half a ton of water entered the collector on top of the airship. Captain Meager, the first officer, went aft to inspect the port fin and found the repair job had stood up quite well, but two long tears of about 20 feet each were made in the lower fabric of the starboard fin. Four hours of repairs with volunteers from the engineers began.

The storm turned south after crossing the St. Lawrence and the airship was advised to return to the river. With the storms in the R100's path to the southwest it was decided to lay to and wait. The navigator, Squadron Leader Johnston, by studying the most concentrated part of the lightning, with the wind on the starboard bow, found how the storms were moving, so the craft could work in behind them. At 12:05 a.m. the R100 officers reported that they were on their way up the St. Lawrence River but would not land until morning because they did not know the location of power lines.

N.S. Norway summed up the tired state of those aboard the R100 upon reaching port: "In the middle of the night, at about two in the morning, the myriad lights of a city showed up ahead of us where Montreal should have been, but in the black sky above these lights, suspended in the night, we saw an enormous fiery cross. I stared at it in consternation till somebody voiced my secret thoughts, and said, 'That's not Montreal. That's the New Jerusalem. This is it, boys.' ".

The steel cross 102 ft. high, 30 ft. across, and 6 ft. thick, had been placed on Mount

R100 sailing up the St. Lawrence River to Montreal, July 31, 1930. PAC PA-119889

Royal's eastern summit in 1924 to ensure its visibility for miles down the St. Lawrence and from the South Shore. Seen from the air on a dark or misty night it appears to float in space.

St. Hubert first saw the airship at 2:30 a.m. when she was just east of Montreal; only her navigation lights could be seen. Word of the R100's sighting in the direction of Mount Royal brought residents tumbling out into the chill of a moonless night. At 2:35 a.m. the piercing ray of the searchlight atop the airport's main hangar caught the British airship for the first time and held her fast while the cheers of thousands and a crescendo of automobile horns provided a rousing welcome.

The glow from the promenade windows, the navigation lights and the steady throb of her motors drew Montrealers back to the airport to join those who had spent an uncomfortable time camped out in surrounding fields.

Showing nothing of the lame duck in her steady motion she headed west, over the Sun Life Assurance building under construction in downtown Montreal. She cruised over the city, the shores of the St. Lawrence and a large portion of the Eastern Townships dodging local storms for 2½ hours.

With streaks of dawn showing in the eastern sky the R100 swung in narrowing circles about the St. Hubert field. A few minutes before 5 o'clock she turned toward the mooring mast. On the passenger platform stood Lieutenant Commander Pressey, the landing officer, and his assistant at the cabinet controlling all the mooring machinery, together with D'Arcy Finn and J.A. Dupont, the English and French radio announcers. The army of camera, movie and newspapermen gathered as close to the mast as the police and the military pickets would permit. All automobile headlights at the airport were ordered extinguished.

The R100 had to valve little hydrogen, having collected almost 5 tons of rain water in her ballast bags. As she slowly approached the mast head from the east at a height of about 500 feet, her 900 foot long mooring cable snaked down in swinging loops to the ground. Her aft engines were put in reverse. Pressey's 14 men, 3 of whom watched the winch drums, allowed the wire to discharge any static electricity in order to avoid a severe shock.

With not a breath of wind the main wire was secured by three of the ground crew and

PAC PMR-73-557

Mooring to the St. Hubert Airport mast at dawn, August 1, 1930, after crossing the Atlantic on her eighth trial flight.

spliced to the ¾-inch cable which had been led from the top of the tower through the mooring arm to a spot determined by the direction of the wind — airships approached a mast head to wind. The ground crew signalled to the ship and the tower. As the engines eased off a little more the cable became taut.

At 5:13 a.m., as the variable speed, electro-hydraulic winch was slowly reeling in the wire, the airship dropped her first water ballast from frame 3 in the bow to keep an even keel. Four bags here, 10 at the control car and 4 at frame 12 each had a capacity of one ton.

The 750 foot starboard yaw guy was paid out 4 minutes later, coupled with a yaw winch cable from the base of the tower and carried to a snatch block on one of the 24 concrete blocks spaced at 15° intervals on the circumference of a circle of 750 foot radius. Reporters and cameramen tried to locate the damaged fin fabric; the looseness of the outer cover was plainly discernible as it rippled in the slipstream of the propellers.

The port yaw guy dropped and was seized by the waiting ground crew and coupled to the mast's second yaw cable. The last engine stopped at 5:25 a.m., but the ship continued to release ballast, some of the water drenching the ground crew and those on the passenger and searchlight platforms.

Even in complete calm it was impossible to bring an airship bow over the centre line of a tall mooring mast and maintain it there while hauling the vessel down several hundred feet. For this reason the 3-wire mooring system and the 2-part telescopic arm or "ram" were developed.

Compressed air pushed out the ram, mounted to swing on gimbals above its centre of gravity, at 6 inch intervals to 7 feet 6½ inches. The mooring rope passed down through the mast to a winch on the ground floor.

Unlike the steam-driven winch at Cardington which was dependent upon two electric motors, which could suddenly fail while collapsing and centralizing the ram, the St. Hubert mast had a main reservoir holding sufficient compressed air for 4 or 5 successive moorings.

Against a background of fading stars, a bluish-gray sky and the golden sun just peering through a saddle in the outline of Mont St. Bruno, the airship descended until the battery of journalists and cameramen could at last plainly read on her silver frame "R-100" and "G-FAAV".

PAC PA-122172

The 60-foot wooden Canadian National Railways Radio tower built on St. Hubert Airport property to report the R100's arrival August 1, 1930. E. Austin Weir, director of the radio department, detailed 12 men to report the airship's progress at Quebec City, Montreal and her mooring.

Broadcasting equipment in the St. Hubert mooring mast for the arrival of the R100. Original plans had Wing Commander Colmore talking briefly from the airship as she approached Montreal.

PAC C-64910

Bell Telephone engineers Lesueur B. Brodie, G. Newmarch and K. Eisner tend radio broadcasting equipment used for the R100's arrival at St. Hubert Airport, August 1, 1930.

Tied in for the government's 1 hour, 20-minute final landing broadcast beginning shortly after 4 a.m. were the Canadian National Railways and Canadian Pacific Railway radio chains, La Presse's CKAC and Canadian Marconi's CFCF in Montreal, and the Columbia Broadcasting System (fed by the CNR) and the National Broadcasting Company in the United States.

Canadian Marconi was to transmit the story over its shortwave beam system for the British Broadcasting Corporation, but adverse atmospheric conditions prevented the graphic account of this historic event from reaching England. Two million British listeners were disappointed, many having waited up a greater part of the night before they learned that the airship had been delayed.

PAC RE-14678

Not yet on Canadian soil, Sir Dennistoun Burney saw St. Hubert and the R100 (here dropping water ballast while mooring) as part of a British-German commercial airship rivalry, with American business the prize. Only the Atlantic and San Francisco-to-Honolulu routes, in his opinion, were self-supporting. The Rocky Mountains ruled out a trans-Canada flight by any airship.

Dungareed mechanics climbed out of the silent engine cars as the ship's dew drop neared the mooring cup. Assistant coxswain T. Hobbs was the first crew member to speak directly to anyone on this side of the Atlantic. When within hailing distance of the mast he called out "hello" from the bow.

Pressey called "Ship secure" at 5:37 a.m. Friday, August 1 as the ram contracted and locked, allowing the airship to swing freely. Immediately the gangway was lowered he went on board. So pleased was he to greet friends from Cardington he had not seen for 6 months, tears streamed down his face.

The mooring in a record 27 minutes versus 35 minutes at Cardington was a signal for rolling cheers from the early risers and those who had remained all night. Automobile horns gave a rousing welcome and police motorcycles within the airport roared into life.

The ground crew immediately attached four ropes from the ship approximately 300 feet, 370 feet, 435 feet and 500 feet from the tower to each of the four stabilizer weights; the one-ton rollers stopped the ship's stern from rising while she followed the wind. The ship's cables were then reeled up on the bow winches. When the airship rode at the mast the dew drop was secured to the mooring cup by locking pins, the mooring wire being replaced by a slipping pennant.

Charles Lindbergh's 33½ hour flight from New York to Paris three years earlier was less time than the R100 spent in passing from the Straits of Belle Isle to the mooring mast. The R100 used more fuel from Belle Isle to Montreal than from Cardington to Belle Isle. The 3,364 nautical miles were covered in 78 hours, 49 minutes including 8 hours for the fin repairs and the subsequent cruising till dawn for an average ground speed of 42 knots. The airship consumed 8,935 gallons (29½ tons) of fuel; 1,505 gallons (5 tons) remained which was enough for another 14 hours at her average rate of consumption. The six engines had run continuously for 23 hours; one engine had run for 59 hours without stopping. Her pressure height on landing was 4,400 feet.

PAC RE-13646

Canada's early morning welcome to the R100 officers and passengers at the base of the St. Hubert mast, August 1, 1930. (From left): Squadron Leader E.L. Johnston, chief navigator (with cigarette); M.A. Giblett, meteorologist; Mayor Camillien Houde of Montreal; Sir Dennistoun Burney, managing director of the Airship Guarantee Company; Wing Commander R.B.B. Colmore, Director of Airship Development, British Air Ministry; Colonel J.L. Ralston, Minister of National Defence; Major G.H. Scott, Assistant Director of Airship Development (Flying), British Air Ministry; Major-General A.G.L. McNaughton, Chief of the General Staff; Captain G.F. Meager, first officer; J.A. Wilson, Controller of Civil Aviation; Squadron Leader R.S. Booth, R100 Captain; Mayor Anatole Lavoie of St. Hubert. Behind Wilson is Nevil Shute Norway, and behind Colmore is Rear Admiral Charles Plunkett, U.S. Navy (Ret'd).

Reporters from newspapers and press agencies were assembling in the mooring mast. They waited there while the customs formalities were carried out. Paul E. Jensen, customs and excise examiner at the airport, entered the airship ahead of the immigration man and the doctor — in a close race — to receive the papers required of every vessel arriving at a Canadian port. The immigration department required a list of passengers and crew stating their country of birth and citizenship. Wing Commander Colmore, Sir Dennistoun Burney and N.S. Norway had all left their passports at home!

J.L. Ralston, the Minister of National Defence, and G.J. Desbarats, the deputy minister, represented the Canadian government. Also in the welcoming party were J.A. Wilson, Controller of Civil Aviation; Major-General A.G.L. McNaughton, Chief of the General Staff; Brigadier-General W.B. King, District Officer Commanding, Military District No. 4 (Montreal); Mayor Camillien Houde of Montreal; Mayor Anatole Lavoie of St. Hubert; R.H. Hadow, Britain's acting High Commissioner to Canada, and several American officers including Rear Admiral William A. Moffett and Lieutenant Commander Charles E. Rosendahl of the U.S. Navy's Bureau of Aeronautics.

The party walked down the stairway inside the mast head to the elevator landing 14 feet below for the one-minute ride to the second floor. Waiting in a room were 20 reporters, who had stood for an hour near the mast during the mooring operation. Burney was the first to enter the room and apologized for being late in arriving at St. Hubert. "I had dinner in London on Monday evening and with only a little luck I should have had dinner in

St. Hubert Airport, August 1930.

Montreal on Thursday. That is the real significance of this flight." The reporters began their questioning of Colmore, Scott, Booth, Johnston and Burney, all showing signs of weariness in their faces.

Squadron Leader Booth stated that much about aerial navigation would be learned from the flight and that with a larger and fast enough airship the Atlantic route could be commercially feasible. In answer to the comparison on everyone's mind to the fin fabric damage occurring to both the R100 and Germany's *Graf Zeppelin* on their maiden Atlantic crossings the ship's captain said the fabric was repaired quite well over the St. Lawrence and he doubted the fabric damage would change the R100's schedule in Canada.

Major Scott, asked to compare his Atlantic flights in the R100 and the R34, smiled and promptly replied that one was very comfortable and one was not. Addressing a small group of reporters in the few minutes that remained Burney explained that the airship had had adverse winds for all but 9 hours of the voyage. To a question of what he did while they sat up all night Burney replied that since it was not necessary to stay up he had gone to bed.

In another group Colmore was also emphasizing the normal routine aboard an airship by saying that meals had been served at proper times although time zones interfered. The previous day watches were set back three hours, which brought meal times "almost painfully close together." He denied he needed to sleep but with his eyes fixed on a "defendu de fumer" notice the 50 cigarettes-a-day Air Ministry official added "but I need a smoke."

A roar of welcome from the large crowd gathered at that early hour greeted the visitors emerging from the building to stand at last on Canadian soil. As pre-arranged, still and movie pictures were now taken and official words briefly spoken. After 80 hours of no smoking the officers and crew readily accepted gift packages of Winchester cigarettes — a moment later recalled in the company's newspaper advertisements.

The night-long vigil was over. Within an hour the quiet of the early dawn had returned to the surrounding fields.

While airport visitors returned to the main entrance or crowded the temporary railway station for tickets and information, reporters proceeded to the press gallery in the main hangar to file the last of their R100 arrival stories.

His Majesty's Airship R100 brought several official letters. British Prime Minister Ramsay MacDonald, in his letter to Prime Minister King, hoped that this flight was the forerunner of a regular airship service because "the growing importance of our political and commercial relations" required faster communications among the members of the Commonwealth.

A message from Lord Stamfordham, King George's private secretary, to Governor-General Viscount Willingdon stated His Majesty was "particularly interested to learn" that the R100 would use the first mooring mast erected by any of the dominions, thus making Canada a partner in "the inauguration of this great experiment."

Lord Thomson, the Secretary of State for Air, wrote Mackenzie King that the mooring mast was one of Canada's major contributions to empire air communications and foresaw, if the experimental program — of which he considered this flight a very important feature — was successful, the building of airships larger and faster than the R100 and the R101. Then Canada's major cities would regard mooring masts as "indispensable" and "leviathans of the air will cross the Atlantic as punctually and safely as ocean liners."

A letter dated July 23 from Lord Thomson to Minister of National Defence Colonel J.L. Ralston stated that Wing Commander Colmore, Director of Airship Development, was the only individual authorized to speak on behalf of the Air Ministry and to settle any questions arising during the visit. Sir William Waterlow, the Lord Mayor of London, wrote Mayor Camillien Houde of Montreal that the building of the civil airport for both airplanes and airships and the erection of the mooring mast would make Montreal "an important link in the chain of imperial air routes."

In a letter to R.H. Hadow, Britain's acting High Commissioner to Canada, J.H. Thomas, Secretary of State for the Dominions, believed that additional communication links between parts of the empire assisted the flow of trade and noted that the airship's voyage came on the eve of the Imperial Conference where the two countries would be discussing their "mutual economic problems and interests." Hadow revealed the letter's contents to the Canadian Press in Montreal because he claimed there was no opportunity for broadcasting it at St. Hubert and he considered the reporters at the airport as "unimportant journalists."

The R100 also carried a letter from Thomas addressed to his son. Hadow, "in view of the historic nature of the occasion", wanted this letter "specially franked" (presumably witnessed) by the airport post office. But as instructions prevented the postal official on duty at that early hour from doing so, Hadow left the letter there with his request in writing.

One unofficial letter was carried by Lieutenant Commander Prentice and dispatched August 1 via New York — a faster method than the St. Lawrence route — to Commander L.D. Mackintosh of the Admiralty's Naval Air Division, Whitehall, London.

The R100's third watch, in Canada since May awaiting her arrival, went on board to relieve the crew, who quickly disembarked to find cigarettes and to cable home or phone friends from the press gallery in the main hangar. The journalists' headquarters were soon dismantled after more than a million words in news stories had been transmitted by wireless, cable and telegraph.

The first concern of N.S. Norway was to have new airship fabric made. In the mast head he met Leonard Hall, an engineer with the third watch and a draftsman while the R100 was under construction, and asked him to make a sketch for the new port fin cover.

Squadron Leader Booth and the third watch began refuelling and gassing the airship while most of the officers were in Montreal where the Department of National Defence had booked rooms in the Mount Royal Hotel. About 50 feet of 12-inch canvas hose leading from the top of the mast through the bow of the ship to a main carried hydrogen from the 50,000-cubic foot-capacity gas holder almost half a mile away on De la Savane Road.

At first there were difficulties with the gas plant, which had a direct line with gas-tight telephones to the control valves on the mast. George Alley, a hydrogen foreman on loan from the Royal Airship Works at £12 per week plus actual expenses had to make alterations to the two generating units, each with a 20,000-cubic foot-capacity per hour, and submitted a list of further modifications before returning home.

Group Captain E.W. Stedman, the Department of National Defence's Chief Aeronautical Engineer, sent for George S. Burrows, manager of the Aviation Department, Canadian Vickers Limited. Burrows and Norway discussed fin cover sketches and went to Vickers in Montreal after lunch. Norway returned to the airship and although he had had no sleep the night before made a second trip to Vickers, the company which had erected the St. Hubert mast. By the time he reached his room in the Mount Royal Hotel for a bath and dinner the airship fabric was half completed.

Even before the airship reached Montreal Rear Admiral Moffett had offered the use of the Lakehurst Naval Air Station if the damaged port fin required more extensive repairs. But Ralston and others explained to Colmore that they were most anxious for the visit to be all-Canadian.

The American government, through the Canadian Legation in Washington, formally invited the R100 to use the Lakehurst, New Jersey facilities. And the Navy Department through the U.S. State Department also extended an invitation to her officers and crew to visit Lakehurst or to inspect the ZRS 4 under construction by the Goodyear-Zeppelin Corporation in Akron, Ohio. The visits were cancelled on August 8, two days after acceptance, because the R100 personnel was needed for installing and doping the new airship fabric and for the Ontario flight preparations.

PAC RE-19795-3

Recorded by Leo LeSieur and his orchestra, the song appeared on two labels: Sterling and Apex.

The one-inch celluloid lapel button is brown on cream with white and blue ribbons.

Proceeds from the sale of this souvenir on St. Hubert Airport property went to charity. Lettering is blue on the one-inch button and the off-white ribbon.

A Montreal Welcome

Greetings to the airship covered Montreal hoardings and shop windows; the Sun Life Assurance Company placed a 150 foot by 30 foot sign "Welcome R-100 to Canada" on its 23-storey building under construction; the airship figured in many newspaper advertisements; poems were composed; flags flew from many roofs and illuminated signs flashed their words of greeting at night.

Souvenirs of the visit sold quickly: lapel buttons, buttons with ribbons, picture postcards of the airship over Montreal and Quebec City and interior views, records and sheet music of the new song "The R-100", and a magazine published by the Advertising Club of Montreal in June and released with an 8 page supplement dated July 28 listing short biographies of the airship's officers and crew.

From the temporary Canadian National Railways station at Guy Street in downtown Montreal trains began their runs at 10 a.m. to the specially-built terminal just outside the main entrance of the airport. Return fare was 50¢ for adults and 25¢ for children. The last train to the airport left at 6 p.m. and left St. Hubert at 10 p.m. The public was admitted to the airport without charge between 10 a.m. and 8 p.m.

For safety reasons the public was not allowed on the airport landing field. Pedestrian traffic routes led from all six gates to the mooring mast where members of the Royal Canadian Dragoons kept the visitors at least 50 feet from the four one-ton rollers attached to the stern of the airship.

Concessionaires dispensed hot dogs, soft drinks and other food. But the Department of National Defence, responsible for the airport and the airship's visit, allowed only R100 souvenirs on behalf of charities to be sold on airport property. In fact, the department stipulated that the 11 souvenir magazine vendors had to be Canadian ex-servicemen.

For the cost of a 5-cent air mail stamp (two cents for postcards) letters dropped in any of the approximately 20 postal boxes on the grounds, and collected every 20 minutes, or mailed from the post office in the administration building, received a black ink rectangular cachet of the R100 at the mooring mast with Mont St. Bruno in the background. American stamp dealer A.C. Roessler of East Orange, New Jersey forged the cachet in order to sell "souvenir" covers, all postmarked August 13, to his clients for 25¢, then 50¢.

With the sunny weather 70,000 people arrived during the day. Despite requests to use the special trains which left downtown Montreal about every 15 minutes most visitors came by car. Eleven nurses and 21 men of the St. John's Ambulance brigade manned stations

PAC C- 37139

The magenta arrival date stamp on a postcard is signed by the airport's postmaster. The Montreal Stencil Works Company made the circular date stamp and probably the black rectangular cachet showing the airship with Mont St. Bruno in the background. American stamp dealer A.C. Roessler forged the cachet to sell "souvenir" covers to his clients.

on the grounds including an emergency tent near the mooring mast. They treated 42 people for sunstroke and epileptic seizures and cared for three children separated from their parents.

Tours of the airship began officially Saturday morning. Reporters were restricted to groups of 15 under the charge of Squadron Leader A.H. Wann. The men were directed to a large wooden box filled with rubber-soled canvas shoes and asked to select a pair. A spark from hard-soled shoes might ignite any leaking hydrogen. Then the visitors took all matches, lighters, pipes, cigarettes and tobacco from their pockets and placed them in a nest of pigeonholes.

Owing to construction differences between the St. Hubert and Cardington masts a two foot gap existed between the three step wooden platform and the R100's gangway. The party walked down the long corridor beneath gas bags hollowed by air pressure at the bottom but still three-quarters full. Upon entering a door the newspapermen could see a steering wheel and elevator controls in the control car below them to their right. A visit to the crew's quarters revealed a room with a table and a gramophone on it at one end and double bunk beds in small cabins on both sides.

Walking up the staircase the groups could believe they were on an ocean liner. The dining saloon was painted white with tables and wicker chairs, the galley at one end, small cabins and promenade decks with brilliant white linen walls. The R100's continuing gassing operation prevented the reporters from visiting the aft part of the airship and the power cars. Wann answered questions about the fuel tanks lining the space below the walkway, which to some moved uneasily, the purpose of valving devices, why garbage was never jettisoned (to maintain equilibrium) and whether passengers could hear the engines, before the tour ended in the dining saloon.

Major-General and Mrs. McNaughton also arrived on Saturday to tour the airship. From the gangway they witnessed Camillien Houde's attempt to enter the vessel. The mayor of Montreal did not let his short legs and the gap prevent him from inspecting the R100. McNaughton admired his spirit as, on command, two of the mayor's police guards picked him up and tossed him onto the gangway.

Among the estimated 3,000 visitors during her stay were government officials, members of Parliament, representatives of aircraft manufacturing and operating companies, and friends of the ship's passengers and crew including chargehand engineer E.J. Stupple who showed Dow Brewery staff over the ship and also the friends he had made in the engineering staff of the Curtiss-Reid Aircraft Company.

Canadian Vickers received the airship fabric order at 2 p.m. Friday and with a double staff worked through the night to complete two panels in 12 hours. One panel was 19 feet by 24 feet and the other approximately 19 feet by 28 feet. Each had 230 stalk patches and lacing eyelets spaced every two inches along the linen taped edges. N.S. Norway returned the next morning to accompany the fabric to the airport before noon but rain prevented putting on the new fin covers.

PAC C-83934

Canadian Vickers aviation department employees apply stalk patches to the new airship fabric, August 2, 1930. Dark-suited gentleman in background is Nevil Shute Norway.

The drizzle also caused the official afternoon reception for His Majesty's Airship R100 to be moved indoors. Planes were moved out of the main hangar, wooden chairs placed and three Union Jacks hung from the girders.

Against the sometimes deafening engine roar of planes Colonel Ralston began his speech by referring humorously to the fact that the people of Canada were "retiring me from circulation" as Minister of National Defence. He was presiding over a ceremony at which "we celebrate the voyage of the world's record airship moored in world's record time of 27 minutes."

This was a great day in the history of aviation in Canada, he said. The country was not abandoning its interest in airplanes for internal development but welcoming the use of airships to improve communications with the rest of the world. He traced Britain's interest in rigid airships; mentioned Mackenzie King's role in supporting the experimental airship program at the 1926 Imperial Conference by constructing a mooring mast; and paid tribute to the airship program personnel the R100 had brought to Canada.

Montreal, according to Ralston, was on "the great circle of international air lines" between the most densely populated areas of North America and Europe and it was possible to visualize the day when St. Hubert "will be the great clearing-house on this continent for

The St. Hubert hangar, with displaced airplanes, where the official reception for the R100 was held minutes before, Saturday afternoon August 2, 1930. The uniforms of the Royal Canadian Mounted Police, the defence forces and the R100 men were a patchwork of scarlet, khaki and blue among the 2,000 spectators. The 33-member Canadian Grenadier Guards band provided music from the rear of the hangar.

Major G.H. Scott told Canada's official R100 reception that experience gained from the Atlantic flight aided the science of airship flying and operation then "still in its infancy." The cylinders of the spare Rolls-Royce Condor IIIB engine peer above its slate-gray packing case.

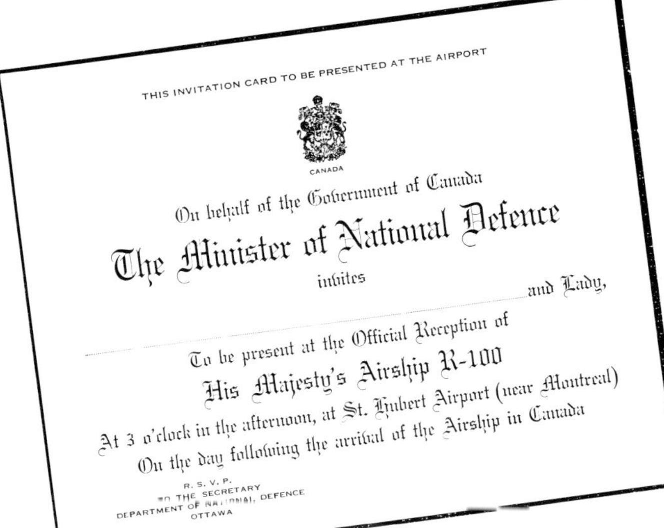

THIS INVITATION CARD TO BE PRESENTED AT THE AIRPORT

CANADA

On behalf of the Government of Canada

The Minister of National Defence

invites

and Lady,

To be present at the Official Reception of

His Majesty's Airship R-100

At 3 o'clock in the afternoon, at St. Hubert Airport (near Montreal)

On the day following the arrival of the Airship in Canada

R. S. V. P.
TO THE SECRETARY
DEPARTMENT OF NATIONAL DEFENCE
OTTAWA

PAC C-83933
This 6¾"x5¼" invitation was also printed in French.

passenger air travel." More rapid transportation and communication would stimulate trade, but there was another reason for Canada's participation in the airship program: "It moves us perceptibly nearer to the British Isles; it warms our hearts with new enthusiasm towards the partner in our successful joint adventure."

Mayor Camillien Houde of Montreal pointed out that in his role of mayor he had welcomed many aviators and long-distance fliers but this was a particularly outstanding occasion. The flight by daring airmen in the face of difficulties had contributed towards "the linking together in closer bonds of unity two most important parts of that great community of nations of which Great Britain is the natural centre and Canada a notable branch."

The last of many speakers, Sir Dennistoun Burney explained that the R100 was the work of many people and he regarded her as the forerunner of a regular Atlantic service. He said that the R100 was an experimental airship but hoped that the next one built could provide 3-day service between England and Montreal and a 2-day return. "It is a bold step to take, but this great empire of ours has been made of bold steps. Faint-hearted policies never did anything."

Sunny 80°F weather on Sunday finally attracted crowds to St. Hubert. Thirty thousand automobiles headed south from Montreal. Twelve-coach trains carrying some of the day's 150,000 visitors left the Guy Street station in the morning as quickly as Montrealers boarded. The Sebastopol Street temporary station began its weekend service to St. Hubert at 2 p.m.

For $5 each way the public could have an aerial view of the British visitor by flying in cabin and open planes from Lasalle Airport on Lower Lachine Road, Montreal to St. Hubert. After gazing upwards at the ship many visitors spent the remainder of their time watching the joy-riding planes taking off and landing. The operators of the nearly two dozen craft, which were in service from the airport's 10 a.m. opening until dusk, were too busy selling tickets and helping passengers in and out of the planes to tally the number carried.

Complaints of poor sales by concession owners and hawkers of hot dogs, soft drinks, badges and photographs of the vessel ended. Restaurant tents outside the airport reported brisk sales. And souvenir magazine vendors were making $6 to $7 a day in commissions. The public now had full access to the grounds.

The Royal Canadian Dragoons were mounted for the first time for their crowd control duties and despite a wind blowing across the field so strong at times that the rollers attached to the airship were lifted several feet into the air the troopers had no problems with the spectators.

Friendly tussles developed between spectators for pieces of the damaged fabric stripped off the fins and allowed to drop to the ground. Quite often the troopers divided the fabric

PAC RE-68-874

Mayor Camillien Houde said the R100 visit was worth $1,000,000 to Montreal. Preparations for a civic welcome began in the mayor's office April 17, 1930 when about 50 representatives of municipalities, service clubs and companies appointed various committees including transportation, publicity and entertainment.

St. Hubert Airport visitors stream past a Canadian National Railways employee from the special trains which ran from downtown Montreal every 15 minutes. The railway provided two tourist sleepers and a lunch counter car for the newspapermen during their long vigil before the R100 arrived. Company president Sir Henry Thornton gave the airship officers and crew unlimited travel privileges.

PAC PMR-73-567

Royal Canadian Dragoon troopers kept crowds 50 feet from the four one-ton stern rollers in case a strong gust of wind caused the R100 to drag the weights along the ground. The rollers (£160) and the mooring tower's main and two guy cables came from the Royal Airship Works, Cardington. Also on duty were the Royal Canadian Regiment, Royal 22e Regiment, RCMP and 100 Montreal policemen.

PAC PA-119888

At the 205-foot St. Hubert mast with Mont St. Bruno in the background. An airship shed would have been necessary for regular transatlantic service. The triangular-shaped property, 90 feet above sea level, with sides of 7,500 feet, 6,700 feet and 7,770 feet, originally contained one stone house, four barns and three small outbuildings.

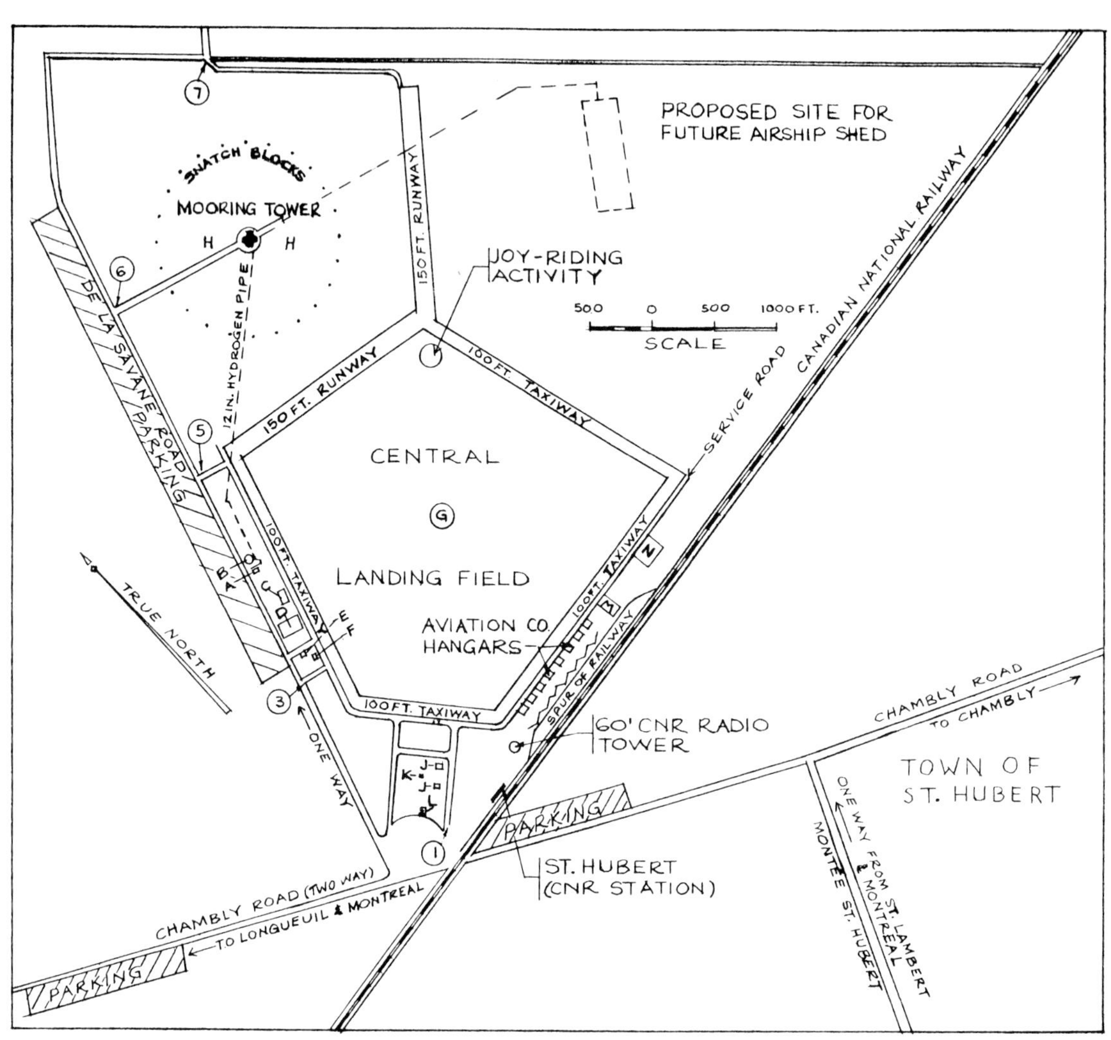

ST. HUBERT AIRPORT, MONTREAL, AT TIME OF R100 VISIT, AUGUST 1930
(NOT ALL STRUCTURES SHOWN)

Legend:

A — Hydrogen Gas Plant
B — Hydrogen Gas Tank
C — Hangar No. 2
D — Hangar No. 3
E — Garage
F — Administration Building
G — 200 ft. Annular Ring
H — Fairlead
J — Houses for Airport Manager & Customs Officer
K — Wireless Station
L — Fire Station and Pump House
M — Imperial Oil Co.
N — Montreal Light Aeroplane Club

Visitor entrances: Gates 1, 3, 5, 6, 7 & detraining to left of aviation co. hangars
~~~~~ CNR excursion station

PAC PMR-73-562

In 13 days 600,000 Canadians and Americans passed through St. Hubert Airport's five entrances and the special railway station (indicated by wavy fencing near bottom right of photograph) to see the British visitor. Joy-riding planes are gathered centre-right in this photograph.

PAC PA-119891

R100 caught in searchlight beams at St. Hubert Airport, Canada's first international airport. This unique photograph was taken by Inspector George Wakeman of the Civil Government Air Operations.

and handed pieces to the visitors. One hawker was charging $1 for a piece. A 40 m.p.h. wind on Sunday again delayed installing the Vickers fabric and doping it with aluminum paint on the outside and red abide on the inside. N.S. Norway decided on Sunday to have Vickers make four fabric panels 20 feet square to carry on the voyage home in case of accidents. Canadian Vickers also carried out repairs to the ship's fuel and oil systems including welding work on the air scoops of the port and starboard power car aft engines and two leaking oil tanks in the same cars.

Visitors to the airship Tuesday afternoon included Lady Perley, wife of Sir George Perley, member of Parliament for Argenteuil and sworn in two days later as Minister without portfolio in Bennett's cabinet, and Miss Mildred Bennett, the prime minister-elect's sister. During their one hour tour they inspected the passenger cabins, electric kitchen, control car and mooring winch. Although Miss Bennett during the inspection commented that the craft seemed fragile she described the R100 to reporters at the base of the mast as "splendid, simply magnificent."

N.S. Norway found the fin repairs were proceeding very slowly although much of the covers were laced in place; the sealing strips would be on the inside. The men, with their only officer on duty in the control car, did not appear to be working too hard in the hot weather. The visits continued. Norway himself gave the captain and officers of the Cunard liner *Ausonia*, which had given the R100 wireless bearings over the Atlantic, a tour of the ship.

On Wednesday, August 6, Major Scott and M.A. Giblett, the R100's meteorologist, made a trip to Toronto. They were welcomed at Union Station by a large crowd including Mayor Bert S. Wemp, Brigadier-General Denis C. Draper, the city's Chief Constable, Captain Earl M. Hand, president of the Toronto Flying Club, H.G. Stapells, acting president of the Empire Club of Canada, and Captain John Barron. The British visitors walked across the street for a civic breakfast in the Royal York Hotel.

In a brief interview Major Scott described the R100's flight as a "de luxe tour" while the R34's Atlantic crossing in 1919 had been a "picnic". The R100 had warm water for shaving, bunks and a dining saloon; the R34 crew slept in hammocks and ate off their knees.

He explained the R100 was not the hundredth airship built for the British government. After the R34 returned from the United States several other airships were launched in that series which ended with the R38. "Everything went flat after that for a while and when the new program was started in 1924 we said 'Let's start at a hundred.' So we did. That's all."

He foresaw regular airship service over the ocean, but not traversing land. "An airship's real job is over the sea. They really are ships . . . and I'm confident of their future."

The temperature was already in the 80's as Scott and Giblett were welcomed to the flag-decked City Hall platform just before 11 a.m. by the whistles and cheers of 5,000 Torontonians. A band from the Mississauga Horse of the Non-Permanent Active Militia played during the reception and three Toronto Flying Club planes flew overhead.

Scott told the crowd and the city council that the R100 should arrive over the city some time on the weekend after her repairs were completed. On behalf of the Corporation of the City of Toronto Wemp presented silver cigarette boxes for Major Scott, Giblett, Captain Meager and Flying Officer Steff. Motorcycle constables escorted the R100 officers back to the Royal York Hotel where they were to be guests of honour at an Empire Club luncheon.

Giblett told the 700 club members and guests how data for airship flying had been gathered from both land and sea since 1925. But an organization was also needed to supply weather information along a proposed route when an airship was about to fly to one of the dominions. Thus he praised the work of John Patterson, Director of Toronto's Meteorological Service of Canada, and his staff at St. Hubert for sending accurate reports before and during the R100's outward voyage.

Pointing to his dark blue uniform which he described as that of an officer of the commercial airship service Major Scott left no doubt of the role of airships in British eyes: "The airship may have some military value; we don't know and don't care. For us, the airship is for commercial transport and for bringing the empire together." He asked his audience for their moral support. "We want you to go forth and say . . . 'I think Canada should insist on the development of these imperial routes and the development of services across the Atlantic.'"

One of four cigarette boxes presented by the city of Toronto August 6, 1930 and treasured by the R100's first officer despite the misspelling of his name.

PAC PA-54760

Harbour Commissioner B.J. Miller introducing M.A. Giblett and Major G.H. Scott at the Toronto Police Amateur Athletic Association annual meet at Hanlan's Point on the Toronto Islands, August 6, 1930. Olympic star Percy Williams won the featured 100-yard invitational race before 6,000 fans in the stadium. Miller represented Toronto at the R100's official St. Hubert reception August 2. Mayor Bert Wemp (holding megaphone) was scheduled to attend a meeting in Green Bay, Wisconsin.

Although only Quebec and Ontario were to see the R100, the Maritimes were also enthusiastic about the airship's visit, as shown by this advertisement of August 6, 1930. Premier J.B. Baxter said that if Saint John, New Brunswick was chosen, his government would "assist generously" with money to erect the dominion's second mooring mast.

While the two R100 officers were in Toronto describing the comforts of airship travel Sir Dennistoun Burney was in Montreal trying to turn his commercial airship scheme into reality.

Burney was the principal speaker at an Advertising Club of Montreal luncheon in the ballroom of the Mount Royal Hotel. He said distance had been killed for the spoken and written word by radio broadcasting, wireless and telegraphy; transportation by water would always be cheaper than any other method for shipping large quantities of merchandise. What was needed was a quicker method of transporting passengers and mail in order to develop the empire's economic and political structure into a reality just as the trunk railways had bound the United States together politically and economically.

The R100 flight demonstrated the feasibility of a transatlantic passenger and mail service. Now Burney was asking Canadians to form a committee and decide to what extent they were prepared to co-operate in such a service.

On Thursday Burney was in Toronto, still trying to get financial backing for his transatlantic airship scheme. Burney informed the Toronto Daily Star in an interview at the Royal York, that he had received a telephone call from New York the previous night from a well-known millionaire offering him $100,000 for a trip in the R100 from Montreal to New York and return.

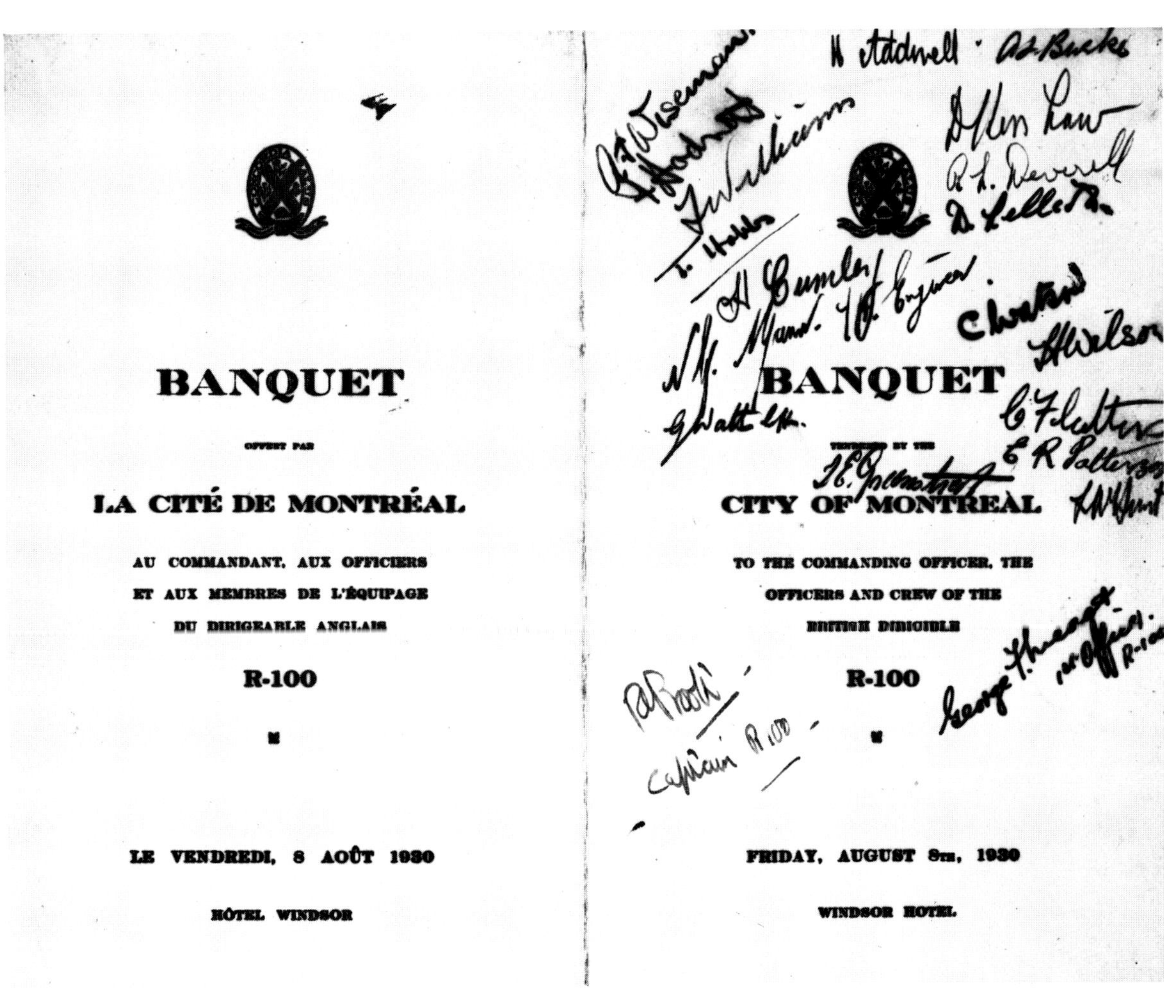

BANQUET

OFFERT PAR

LA CITÉ DE MONTRÉAL

AU COMMANDANT, AUX OFFICIERS
ET AUX MEMBRES DE L'ÉQUIPAGE
DU DIRIGEABLE ANGLAIS

R-100

LE VENDREDI, 8 AOÛT 1930

HÔTEL WINDSOR

BANQUET

TENDERED BY THE

CITY OF MONTREAL

TO THE COMMANDING OFFICER, THE
OFFICERS AND CREW OF THE
BRITISH DIRIGIBLE

R-100

FRIDAY, AUGUST 8TH, 1930

WINDSOR HOTEL.

PAC C-37140

Speeches in French and English praising the R100 officers and crew and song-singing led by Mayor Camillien Houde highlighted Montreal's civic banquet August 8, 1930. The menu for the 800 representatives of the church, business, politics and the Colombian and German consulates included consommé Madrilene, filet Doré Meunière, chicken, new potatoes Persilles and bombé glacée R100.

Offering to go as high as $200,000 through a New York agent, wealthy 24-year-old American oil man and movie producer Howard Hughes wanted the flight to coincide with the opening of his World War I aviation picture "Hell's Angels" at two New York theatres August 13. Publicity plans included airplanes carrying R100 crew members from Lakehurst landing on the Hudson River. The men would then proceed to the two Broadway theatres for the movie's premiere. "I told him that the R100 was at the present time a government project and not yet operating commercially. It was a shame to throw away money like that for it wouldn't take many of those customers to put the line on a paying basis."

The R100 crew which normally stood 4-hour watches was called out of their second-floor mast quarters around 3 o'clock Friday morning for an electrical storm which ended Montreal's hot and sultry weather. It rained so hard that 13 tons of water ballast were dropped but when the sun rose the ship almost stood on end before sufficient water was piped on board to recover an even keel.

A meeting at St. Hubert Airport Saturday between Colmore and J.A. Wilson confirmed that a local flight would begin Sunday evening August 10. Squadron Leader Booth did not believe a reliable southern Ontario schedule could be given in advance so the government would send messages to Ottawa and Toronto radio stations during the flight advising them when to expect the airship.

Major G.H. Scott (left) in the R100 control car over southern Ontario

The Local Flight

A stiff northwesterly breeze swung the R100 at the mast; she in turn dragged her huge rollers along the ground. The breeze subsided early Sunday evening but bumpy weather was predicted for the journey. The ship was closed to visitors in the afternoon to complete arrangements for the 24-hour flight.

The passengers were senior military men, officials of the Department of National Defence and two newspapermen: Major-General A.G.L. McNaughton, Chief of the General Staff; Commodore Walter Hose, Chief of the Naval Staff; Group Captain E.W. Stedman, Chief Aeronautical Engineer, Department of National Defence; Group Captain J. Lindsay Gordon, Director of Civil Government Air Operations; Wing Commander L.S. Breadner, Director, RCAF; Commander C.P. Edwards, Director of the Radio Branch, Department of Marine and Fisheries; Captain Frank Badgley, Director, Canadian Government Motion Picture Bureau, Department of Trade and Commerce; Squadron Leader J.H. Tudhope, Superintendent of Airways, Department of National Defence; Brigadier-General W.B. King, District Officer Commanding, Military District No. 4 (Montreal); Lieutenant-Colonel W. Arthur Steel, and Captain W.L. Laurie, both of the Royal Canadian Corps of Signals; Flight Lieutenant Frank S. Coghill, Operations and Intelligence, RCAF; R. de B. Corriveau, Assistant Chief Engineer, Department of Public Works; W.A. Lawrence, press liaison officer, Department of National Defence; Captain Frank Tyrell, official photographer from the Canadian Government Motion Picture Bureau; Major William Baty, Jr., Camp Commandant, St. Hubert Airport Camp; George MacDonald of Montreal, Eastern District Manager of the Canadian Press; and a Reuters representative.

In the administration building these 18 Canadian guests signed indemnity and waiver forms arranged between the Department of National Defence and the British Air Ministry. No signatory, nor his heirs or executors, could bring legal action against the King or his officers and servants or the Air Council if he was injured or died while a passenger on His Majesty's Airship R100.

With complete co-operation of the crowds the Royal Canadian Dragoons troopers cleared a 300-yard circle around the mast. At 5:35 p.m. Squadron Leader Booth, shouting through a megaphone from the control car, ordered the six-man ground crew to cast off the ropes attached to one of the stern stabilizing weights just before Major Scott and navigator Squadron Leader Johnston, with charts and instruments under his arm, boarded the ship. The ropes swung up and were hauled into the vessel as she swayed slowly from west to north and back again while her stern rose and fell lifting one of the rollers 20 feet into the air.

PAC RE-13637

Arriving from Ottawa for R100's flight over southern Ontario: (From left) Group Captain E.W. Stedman, Chief Aeronautical Engineer; Commodore Walter Hose, Chief of the Naval Staff; Major-General A.G.L. McNaughton, Chief of the General Staff; Group Captain J. Lindsay Gordon, Director of Civil Government Air Operations; Wing Commander L.S. Breadner, Director, RCAF; Commander C.P. Edwards, Director, Radio Branch, Department of Marine and Fisheries; and Captain Frank Badgley, Director, Canadian Government Motion Picture Bureau, Department of Trade and Commerce.

Departing five minutes later from the administration building motorcycle policemen with sirens wailing escorted the five automobiles carrying R.B.B. Colmore, Director of Airship Development, and the R100's guests to the mooring mast. In the lead car were Mayor Camillien Houde of Montreal and his wife who had come to wish Godspeed to the passengers.

To the newspapermen and movie cameramen gathered before the tower entrance Colmore briefly stated the itinerary the R100 would follow: she would circle Ottawa that evening, fly over Toronto at 9 o'clock in the morning before heading for Hamilton and Niagara Falls and then visit Kingston on her way back to Montreal. The passengers paused long enough for photographs and then proceeded to the mast elevator at 6 p.m.

The last weather report from the airport meteorological office was "wind, west-southwest, 17 miles per hour, cloudy, strato-cumulus cloud, visibility excellent, temperature 65 degrees F."

Water ballast was dropped to trim the ship as two more stern rollers were uncoupled. The crowds closely watched movement within the leviathan, and heard bells in the power cars acknowledging commands as three engines started up. A policeman stood on guard on the passenger platform of the mast, while above him four men on top of the mast head awaited the release of the mooring eye from the cup. At 6:17 p.m., the gangway was shipped. Booth leaned from the control car window and signalled the release of the fourth stabilizing weight.

One minute later cheers rose from the crowd of 200,000 gathered in a horseshoe which stretched two miles from tip to tip. The R100 ruled the sky. All other aircraft were forbidden to be in the air 15 minutes before her departure. The R100, whose passengers waved in return to the good wishes of the crowd, slowly withdrew from the orange and black mooring mast head and stopped 200 yards from the mast before reversing her Rolls-Royce engines. She briefly sailed towards the descending sun then headed in a northwesterly direction, climbing to 1,200 feet.

The flight would show the R100 to many cities and towns and would allow her Canadian guests on board to experience the comfort of airship travel. It was also an opportunity to test her fin repairs. There was no thought of setting a speed record.

But first Squadron Leader Booth turned south to avoid two squalls he saw in his path. Montrealers saw the airship in the distance, but many thousands on top of Mount Royal and those who had occupied other vantage points in the hope of a close-up view of the R100 were disappointed as she headed for the south shore of the St. Lawrence River and Lac Saint-Louis before setting a course for Ottawa.

Scarcely more than five minutes after her departure it began to rain. There was a rush to automobiles and to the trains returning to Montreal's Guy Street station. The deluge ended one of the concessionaires' better days. The R100 missed the bad weather. The passengers remarked on the airship's apparent lack of motion and looking back to the storm area they took the gorgeous rainbow they saw as an auspicious sign for the journey ahead.

The R100 passed over Oka, Lac des Deux Montagnes and Hudson Heights before crossing into Ontario where she reached Plantagenet, 35 miles from Ottawa, at 9:10 p.m.

Everyone enjoyed dinner, served by stewards. After partaking of chicken and ham, salad, bread, butter, cheese, biscuits and coffee, the passengers had to settle for gum in place of the customary after-dinner cigar or cigarette. Some spent the evening playing bridge while others watched the setting sun or the scenery passing beneath them.

While the passengers were spending a quiet evening the citizens of Ottawa were assembling to greet the British liner. They began arriving on the lawns of Parliament Hill

PAC C-47144

Tables set in R100's dining saloon in this photograph taken by local flight passenger Lieutenant-Colonel W.A. Steel.

around 7:30 p.m. with blankets on their arms until their number reached 20,000. Thousands more crammed the nearby vantage points of Major's Hill Park and Nepean Point expecting the R100 shortly after 8 o'clock. The 53-bell carillon of the Parliament building's Peace Tower provided music from 9 o'clock to 9:45 while the air grew chill as a wind swept in from the Laurentian Hills. At 9:50 the airship was first sighted from the Peace Tower as a black dot on the southern horizon. As she drew closer the light of the brilliant silver moon gave her a gunmetal colour.

The cheering began when a dozen searchlights on the East and West Blocks of government buildings on Parliament Hill picked up the letters R-100 which were clearly visible even at 2,000 feet. The new fabric on the port fin could also be discerned. At that moment the carillon rang out "Rule Britannia" and "O Canada." The blare of automobile horns and railroad locomotive whistles as the airship came into view temporarily drowned out the arrival commentary by Canadian National Railways' radio station CNRO.

This night also marked the first time in Canada home listeners could hear a two-way communication between the ground and a moving aircraft. CNRO broadcast on one wavelength and the Ottawa station of the Royal Canadian Corps of Signals received the airship's signal on 1400 metres. Both outputs were passed through a mixing panel in the radio station in the Jackson Building at Slater and Bank Streets. From the tower of the Chateau Laurier Hotel, beside Parliament Hill, Prime Minister Bennett and Mayor Frank Plant of Ottawa exchanged greetings with Colmore and Major Scott.

While over the capital the airship received requests from the Commandant of the Connaught Rifle Ranges, where the Dominion of Canada Rifle Association was assembled for a meet beginning the next morning, and from the mayor of Carleton Place to visit them while en route to Toronto. Colmore informed his radio listeners of these two additions to the R100's itinerary.

The man responsible for accepting the mooring mast for Canada seemed somewhat bitter at the thought of other politicians getting the attention on this historic evening. Mackenzie King described this radio exchange, which was at times indistinct, as "a banal sort of performance, and not at all suitable for Sunday night." Wondering why the airship could not have flown over Kingsmere, his summer home across the Ottawa River in the Gatineau Hills, King assumed G.J. Desbarats, the Deputy Minister of the Department of National Defence, had not relayed the suggestion to the ship's officers. "No one seemed to think of arranging for this, though to me more than anyone the whole venture so far as Canada was due. They might easily have come out over King's Mountain and made the trip especially, but yesterday Prime Minister, first in everything, today nothing."

The giant airship, with her red, green and white lights, passed over the Parliament building at 10:09 p.m. Ten minutes later she visited Rideau Hall, the residence of the governor-general, in the Rockcliffe area of the city. Turning westward the airship flew over the city of Hull on the Quebec side of the Ottawa River before visiting the southern end of Ottawa. Heading north she passed low over the Peace Tower to the cheers of the enthusiastic crowds below and dipped her bow in honour of Canada's war dead. The carillon in the tower now played "God Save the King." The Connaught Rifle Ranges were the R100's last place of call before beginning her zig-zag route through the night to Toronto. Most of the passengers were now ready for bed.

The southern Ontario radio listeners who had heard the R100's triumphant visit to the capital hoped to see the airship as she headed through the cloudy sky to Toronto. Leaving Ottawa shortly after 11 p.m. the airship passed over the Rideau Lakes, whose moonlit waters moved one passenger to describe the lakes as "seeing jewels in their best setting." The first town reached after Ottawa was Carleton Place.

As the R100 approached Smiths Falls town employees climbed the water tower to turn on the aerial beacon. Thus the residents saw her caught between the beacon light and a full moon as she flew over the main street. After her midnight call on Smiths Falls the airship set a course for the Thousand Islands just east of Gananoque. Kingston was reached at 1 a.m. The citizens of Belleville saw her pass 10 miles south of the city heading in a northwest direction.

At 3:10 a.m. the R100's engines announced her arrival to the light sleepers of

Peterborough. Now 65 miles northeast of Toronto she turned to Lake Ontario, passed over Oshawa at 4:10 a.m. and paid an early morning visit to the Queen City. The R100 approached Toronto harbour from the Eastern Gap and many of her passengers were woken at 4:45 a.m. by boat whistles as she sailed over Bay Street at 2,000 feet. At home one Globe newspaperman, awakened from a light sleep, described the R100 as "a delicate silver fish against the deep rich blue of the sky with the coloured port and starboard lights like rubies and emeralds amid the scintillating diamonds of the stars!" But one Torontonian on the streetcar admitted that the rhythmic beat of the engines did not induce him to search the sky for the visitor. "I stayed in bed and reflected that if the airship is a success there will be plenty more of them to see, and if it isn't a success it isn't worth catching a cold about."

Over City Hall the airship turned west along Queen Street then moved slowly across the city in a northwesterly direction before turning south for Niagara Falls. At 5:30 a.m. the airship made the first of her four passes that morning over Niagara Camp, the Royal Canadian Dragoons "B" Squadron's summer camp at Niagara-on-the-Lake.

After visiting the falls she flew over Niagara Falls, New York before returning to Canada to circle the nearby city of St. Catharines and the Welland Canal and went on to Hamilton where at 6:30 a.m. the residents were treated to the beautiful sight of the airship's cover glistening in the early morning sun.

Many Hamiltonians were later disappointed in not seeing the airship which was expected at 10 a.m. Approaching from the lake she circled the city and Burlington Bay (Hamilton harbour) clockwise while cruising at approximately 40 m.p.h. at 1,500 feet. She passed almost directly over the municipal airfield. Local aviators would have provided a guard of honour but were forbidden by government regulations from doing so.

After about 15 minutes she headed back to Toronto through a huge bank of white clouds and after reaching the Western Gap of Toronto harbour at about 7 a.m. turned about to pay another visit to St. Catharines shortly after 8 a.m. Returning to Niagara Falls she sailed up the Niagara River to the famous cataract where her passengers enjoyed breakfast 1,500 feet above the falls.

The ship flew over the Peace Bridge and again circled nearby Niagara Falls, New York. She then sailed back down the river and out into Lake Ontario to keep her appointment with Toronto. Boats on the lake bound for Niagara Falls whistled their greetings.

The airship, powered only by her three pusher airscrew engines, again approached the downtown Toronto area from the Eastern Gap. This time Torontonians gave the British visitor a royal welcome.

Downtown rooftops were crowded by office workers. A huge Union Jack, measuring 24 feet x 12 feet, and a "Welcome R-100" sign painted in letters 15 feet high greeted the airship as she passed over City Hall at 9:18 a.m. Well-wishers at this vantage point thought she was greeting the crowd by dipping her bow. Unknown to them the airship was encountering very bumpy air over the city.

The airship passed quite close over the top of the nearby 34-storey Canadian Bank of Commerce building — the tallest in the Commonwealth — before heading north to the midtown area of the city. Here the passengers saw the Ontario Parliament building and the University of Toronto campus.

On one of her circuits of the city she flew so low over Stanley Barracks, home of the Royal Canadian Dragoons "B" Squadron, that passenger Major William Baty later told his groom he could see dirt on his horse! The R-100 Restaurant on Yonge Street near Briar Hill Avenue opened that day and kept the name for 8 years.

The R100 proceeded westward as far as the Humber River before heading south to the somewhat smoother air over Lake Ontario. Crossing the Toronto waterfront she left the bright sun and blue sky behind and headed into the mist and grayness of the eastern sky. Although the Queen City saw her for only an hour the Toronto Daily Star summed up the excitement with its headline "Industry At Standstill As R-100 Visits City."

The Canadian Forum of September 1930 perhaps better explains why Toronto eyes were skyward: "On her western course she passed by low at night, vast, majestic, gold-washed by the moonlight against a serene dark sky, her lights as steady and assuring as those of a liner at sea, the drone of her engines implicit of conscious power. On her return

Horseshoe Falls photographed from the Canadian side of the Niagara River, Monday morning August 11, 1930. The R100 had permission to fly over Niagara Falls, New York.

Torontonians crowded rooftops as the R100 approached the city from the Eastern Gap, Monday, August 11, 1930. The Canadian Bank of Commerce building, in centre, was then the tallest building in the Commonwealth. This view from City Hall was probably taken by Arthur Goss, the city's official photographer.

PAC RE-13643

Local flight passengers on the promenade deck, August 11, 1930. Facing camera is Lieutenant-Colonel W. Arthur Steel, Chief Technical Officer of the Royal Canadian Corps of Signals, who had commanded the communications centre in St. Hubert's administration building as the R100 approached Montreal July 31. Steel joined the Canadian Radio Broadcasting Commission in 1932 as one of its first three commissioners.

PAC PA-111268

The Toronto Islands and downtown area from the R100 on her visit to the Queen City, August 11, 1930. The Canadian Bank of Commerce building is in the centre and the City Hall clock tower is in lower right of photograph.

PAC PA-54756

Over downtown Toronto on her second visit to Ontario's capital in five hours, Monday, August 11, 1930. Looking northeast from Wellington Street East at Yonge Street St. James' Cathedral is partially hidden in background.

Over a Toronto landmark under construction, August 11, 1930 — the Canada Life Assurance building with its weather beacon on University Avenue.

we saw her in the pride of day, as much at home in her element and as beautiful as a full-rigged ship on blue water. A sight to lift a man's heart. There may be a hundred reasons why these ships of the air cannot be a commercial success: what does that matter! The machine age, which has filled our lives with noise and stinks and soul-cramping ugliness, can also give us things as poetically romantic as the R-100. Whether we can afford them or not, let us have all of them we can get, say we."

A "Welcome R-100" sign 15 feet high on Toronto's City Hall driveway greeted the airship August 11, 1930. Citizens returned to the cenotaph (on the left) October 11 for an R101 memorial service.

The R100 headed for the southern shore of Lake Ontario where the air was less bumpy. At 11:40 she was about 15 miles from the American shore doing 65 knots with four engines running. The aroma of grilling chops filled the passenger compartment. Stewards were setting tables and arranging flowers on them in preparation for lunch.

Half an hour later the British visitor was off Rochester, New York at an altitude of about 1,000 feet and proceeded to follow the American shoreline to Oswego. When almost opposite the mouth of the Oswego River the airship turned due north for Kingston, Ontario.

While over the lake Major-General A.G.L. McNaughton, Chief of the General Staff, accepted an invitation to tour the ship. To reach the aft observation platform meant pushing against gas bags with his back in order to gain two steps on the 80 foot ladder at frame 12. Upon seeing the outer cover he declined to walk along the 9 inch-wide catwalk with only a rope to grab in an emergency and went below to continue his technical inspection.

In the daylight Kingston residents had a better chance to see the airship. Many of the passengers had sent souvenir messages to friends in towns in the ship's path while she exchanged greetings with radio stations in Toronto, Camp Borden and now Kingston. The passengers identified the campus of Queen's University and other areas of the city and, once past the harbour, the Royal Military College and nearby Fort Frederick and Old Fort Henry.

Thousands of summer visitors and local residents crowded the river shore at

Gananoque to see the giant airship on her second visit in 13 hours. The ringing of an old fire alarm bell on the roof of Brockville's United Counties building gave residents time to find vantage points before the R100 passed over the American shore opposite the city at 2:12 p.m. Bells and whistles played a similar role in Prescott where people gathered in the streets, on rooftops and on the docks under a cloudy sky and light rain to see the airship over the St. Lawrence River.

PAC PA-111269

Following the St. Lawrence through the Thousand Islands, August 11, 1930. Ontario is on the left and New York State on the right.

The R100 flew directly over Cornwall as part of the control car's visual indicator experiment using the directive radio beacon sent out by the Canadian Marconi Company's test station at Laprairie, Quebec six miles from St. Hubert Airport. The radio signal the airship was to follow to Ottawa ended 10 miles before the capital, but her officers found the signal directed through Cornwall quite strong and calibrated correctly and followed it almost to St. Hubert.

Around 4 p.m. the R100 appeared over Lac Saint-Louis, part of the St. Lawrence River at the southwestern shore of Montreal Island. She then sailed over the Lachine Rapids.

Office workers crowded Montreal rooftops as the airship on her second pass over the city approached from the west, crossed St. James Street and then turned north to fly over McGill University and Mount Royal. As she completed her circuit the vessel appeared to spectators in Dominion Square to be barely above the treetops. She continued in an eastward direction for a short distance before turning south towards St. Hubert. Thinking that the R100 was intending to moor many people began heading home.

St. Hubert, which along with Ottawa had been in constant touch with the airship, kept the R100's navigator informed about mooring conditions so he could determine how much time to allow for sightseeing over Montreal. The main reason for continuing to circle the city for two hours was to avoid local thunderstorms.

While the vessel was dodging the storms the reduction gear of the forward starboard engine broke up. Pieces of the casing were thrown by the propeller up into the airship, not only causing a tear in the outer cover while the propeller spun into the St. Lawrence River

R-100

The airship nearing the end of her local flight, August 11, 1930. Below her in Montreal's Victoria Basin are three white-hulled St. Lawrence River cruise ships of the Canada Steamship Lines. The Harbour Bridge, now Jacques-Cartier Bridge, crosses St. Helen's Island, subsequent site of Expo 67.

A postcard of the airship over Montreal, August 11, 1930, and the Sun Life Assurance building's R100 welcome sign.

PAC PA-111271

Downtown Montreal from the R100, August 11, 1930. From left to centre of photograph: Windsor Station, St. James' Cathedral (now Cathedral of Mary Queen of the World), facing Dominion Square, and, under construction, the Sun Life Assurance building.

While circling Montreal, August 11, 1930, after her visit to Ontario, the R100 flew over the Beaver Hall Hill headquarters of the Bell Telephone Company of Canada, now Bell Canada.

but, more importantly, making a hole half the circumference of the tube in the base boom on the transverse girder and main joint at frame 9. Also, while she was cruising over the city the magneto of the starboard aft engine failed, throwing out the timing. As the engine could not then be run in reverse it was stopped.

Despite these engine problems, of which the passengers were unaware, the ship was moored in a record time of 25 minutes at 8:15 p.m. In the air for 25 hours, 57 minutes the R100 covered the 805 miles using 2,650 gallons of fuel at the rate of one-third of a ton per hour. Her pressure height leaving the day before was 3,000 feet and 4,500 feet on her return.

A spare Rolls-Royce Condor IIIB engine sent from England was available to replace the damaged forward starboard engine. But unless the airship was released from the mooring cup and drawn down to the ground by a large landing party, it could not be installed. The specially designed two-legged derrick, which could be fitted in the centre section of an engine car allowing men in the car to hoist engines in and out with a chain purchase while the airship rode at a high mast, was still in England. N.S. Norway found it irritating that an "organizational failure" at the Royal Airship Works at Cardington prevented this special equipment from being in Montreal when needed.

The damaged magneto and propeller were replaced and the fabric torn by pieces of the reduction gear casing was repaired from inside the hull. Also replaced was the ship's main wire which had been pulled in two after coupling to the mooring arm. Lieutenant Commander Pressey, in charge of the mast, concluded that the wire must have been kinked because there could not have been more than a 15-ton pull on it.

R100 engineer Leonard Hall went to Canadian Vickers to get a patch made for the base boom of the transverse girder. Norway, after inspecting the damage with chief engineer W.Y. Angus and Frank McWade, the Royal Airship Works' resident inspector of the Aeronautical Inspection Department, decided to put on a double hole patch plate with two 1-inch holes to be riveted with fitted bolts. But after engineer D. Lelliott phoned Norway

R-100 Flies With Imperial Gasoline

Fuel for her voyage over Eastern Canada and for her return trip to England has been furnished the R-100 by Imperial Oil Limited.

The more exacting the requirements the more certain is the choice of Imperial Oil products which are guaranteed by the unequalled experience and equipment of Canada's largest producers, refiners, distributors and marketers of petroleum and petroleum products.

IMPERIAL OIL LIMITED

Serving Canada for 50 years

6 Refineries   1900 Branch Warehouses   Thousands of Dealers

Three railway tank cars, each with a capacity of 7000 gallons, were required to transport the R-100's fuel supply from Imperial's Montreal East refinery to St. Hubert Airport.

Imperial Oil sold the government 20,000 gallons of gasoline (20% benzol mixture) in May 1930 at 28½¢ a gallon and bought back the remaining 9,448 gallons in October 1931 at 15¢ a gallon less freight and handling charges. It is unlikely the company recovered its expenses in matching British specifications for the airship fuel; some of the ingredients were not available in Canada. One R100 engineer considered the fuel mixture superior to the British fuel.

A SAFE JOURNEY OVER AND A TRIUMPHANT RETURN!

Air-Worthiness of the R100 was Proven on CASTROL

AGAIN it has been done! The vast Atlantic has been bridged by air! New history has been made . . . and Castrol helped to make it!

Prior to the time the R-100 was taken over by the British Air Ministry, its engines were put to the most gruelling trials. All lubricants were available. CASTROL, alone, was chosen. Every mile was flown with a *Standard* grade of this World Famous Motor Oil!

For both the Atlantic Crossing and the Return Flight, the Air Ministry specified their Special Castor treated grade . . . in accordance with Air Ministry practice, the fuel also being a Special mixture blended to conform with Air Ministry specifications.

In the great achievements of Yesterday and amid the amazing advances of TODAY, CASTROL is supreme among the lubricants of all nations. In the stupendous undertaking of Tomorrow, Castrol will be playing its part.

Produced by

C. C. WAKEFIELD & CO. LIMITED

"The All-British Firm"

Montreal - Toronto - Winnipeg - Regina - Vancouver

Head Office—London, Eng.

What About YOU?

Experience is the greatest teacher. The stranger is not to be entrusted with the same confidence we place in an old and proven friend. Adopt Castrol for your car. Profit by its Economy . . . and learn what Super-Lubrication can mean to security, satisfaction and peace of mind.

The R34 Used CASTROL!

Eleven years ago the British built R34 made the first return flight across the Atlantic. A standard grade of CASTROL was the lubricant.

More Outstanding Achievements To Its Credit Than All Other Oils Combined

around 7 o'clock in the evening to say the patch would not fit the slightly bent tube Norway agreed to use Hall's design with small modifications.

On Monday, August 11 G.J. Desbarats, Deputy Minister of the Department of National Defence, sent a tentative list of the return flight passengers to A.W. Merriam, the prime minister's private secretary, and noted that although the R100 was scheduled to depart Wednesday she might leave Tuesday night. The list was returned to the department at 2 o'clock in the morning with the only change being the deletion of P.E. Boivin, mayor of Granby, Quebec, and member for Shefford until the general election.

Although the R100 would soon leave for home Montrealers learned in the press that in September they might see the *Los Angeles* which had verbally accepted an invitation to visit St. Hubert. When she said no several weeks later the mast was shut down for the winter.

At St. Hubert the airship's navigating officers were closely inspecting weather charts and navigation maps with the government meteorologists to determine probable weather conditions and favourable routes for the voyage home. Government officials from Ottawa toured the airship and her mechanics ran the five working engines in turn for about an hour at varying speeds in the afternoon.

If there was a calm wind that night Colmore proposed drawing the R100 down to the ground and changing the damaged engine using a small crane. Scott, Booth and Norway were against this course of action claiming there was considerable risk of damage to the power cars and the rest of the ship as the vessel was let up to the mooring cup after the forward starboard engine and substructure were removed and hauled down again to fit the new engine.

It was definitely agreed in the evening to leave the disabled engine in place and return home on the remaining five engines, the deciding factor being the prevailing westerly winds across the North Atlantic.

PROEM

These lines are inscribed to the gallant officers and adventurous gentlemen whose genius conceived and courage navigated the British Airship R-100 from England to Canada August 1930.

August 9th, 1930.

The Shade of the British Airship R-100

(By Fred Rowlett)

"Queen of the Air, I soar serene,
Scorning the trodden ways of men.
Nor bound by Earth, nor slave to Tide,
I glide o'er mountain, vale and fen;
Soaring o'er all the Western Sea,
Bouyant, and beautiful, and free!

Westward I turn my radiant face,
Naked I ride Celestial Air;
Westward, nor falter, dusk nor dawn,
To greet thee, Canada, most fair;
A glorious land that is to be,
A Nation rich, and strong, and free;
A Silver Star, upon an
Azure sea."

Montreal to Cardington

On Tuesday evening these brief words appeared under "arrivals and departures" on the bulletin board in the administration building: H.M. Airship R100 for Cardington, England, 13 August, 10 p.m. The R100 crew members, who had learned that day which 9 of their number would be returning to England by sea, held a farewell party in a room in the administration building. Squadron Leader Wann was to be in charge of the watch, but instead made the round trip with the 11-man relief watch in Canada for 11 weeks.

Several officers of the Department of National Defence and Post Office Department officials applied to make the return trip. But the Department of National Defence decided to send only one technical officer from the Canadian government: their Chief Aeronautical Engineer, Group Captain E.W. Stedman, who as far back as April had requested permission to accompany the airship from England to Montreal; he was also a passenger two days before on the local flight.

The Prime Minister's Office nominated Jacques-Narcisse Cartier, a Montreal newspaperman (La Presse) with an historic surname, for one of the 10 passages the Air Ministry had allotted Canada for the flight. As he wrote Lord Thomson on August 13, Prime Minister Bennett expected Cartier "will undoubtedly do much to create a favourable public opinion of the commercial use of airships" after his return.

His fellow reporters would be Don C. Brown of the Mail and Empire (Toronto) and also representing the Manitoba Free Press (Winnipeg), J. Fergus Grant, aviation and marine editor of the Gazette (Montreal) and also representing the New York Times, C.H.J. "Jerry" Snider of the Evening Telegram (Toronto) and also representing the North American Newspaper Alliance, and Thomas Wayling of the Toronto Daily Star and also representing the Southam newspapers and the Times of London.

The Air Ministry filled its quota of five passengers with Wing Commander L.J.E. Twistleton-Wykeham-Fiennes, the British air attaché in Washington, and four newsmen: Giorney Bolton of the Yorkshire Post and the Times of India, E.W. Grange of Reuters Cable Agency, Ottawa, and also representing the Associated Press, Glyn Jones of the Central News and Montague Slater of the London Daily Telegraph.

Official mail consisted of letters from Prime Minister Bennett to Prime Minister Ramsay MacDonald, Bennett to Secretary of State for Air Lord Thomson, Minister of National Defence Donald Sutherland to Lord Thomson, the British High Commissioner's Office to the Secretary of State for the Dominions J.H. Thomas and Mayor Camillien Houde of Montreal to Sir William Waterlow, the Lord Mayor of London.

At the Royal Empire Society luncheon in the Windsor Hotel August 7 Squadron Leader

Booth consented to carry a photograph of the head table guests and a letter written by Lieutenant-Colonel Charles Adams of the Montreal branch for the Royal Empire Society in London. Sam Bronfman of the Distillers Corporation handed Squadron Leader Johnston at a joint company-city luncheon August 13 a letter for William H. Ross, president of the Distillers Corporation of Great Britain.

The known unofficial mail consists of 26 air mail covers American stamp dealer A.C. Roessler of East Orange, New Jersey got a crew member to carry for him; a picture postcard cancelled at St. Hubert Airport August 13 of Squadron Leader Booth and the R100, written over mid-Atlantic by passenger Thomas Wayling to Mrs. Norma Muir of Toronto and posted in London August 16; and a postcard carried by engineer Cyril Watson, in Canada since May with the third watch, and posted in Bedford August 16 to Lesueur Brodie of Montreal, a Bell Telephone engineer on communications duty at St. Hubert Airport during the R100's stay.

N.S. Norway made certain that the four 20 feet x 21 feet spare panels of airship fabric were on board. Also carried for emergency use were two aluminum oil tanks and 170 linen and leather stalk patches — all material from Canadian Vickers.

The day before the airship's departure the first of her special cargo had arrived from the Niagara peninsula. Three crates of peaches, courtesy of Ontario Premier G. Howard Ferguson, were flown to St. Hubert by two members of the St. Catharines Flying Club. Two of the crates were for the R100's officers and the third containing 12 dozen peaches was for the Prince of Wales. By air on the 13th the governor-general sent a box of peonies cut by his wife that afternoon at Rideau Hall as a gift for King George and Queen Mary.

The passengers assembled at the airport's administration office at 8 p.m. Here they were weighed, along with their luggage, and inspected for matches and hard-soled shoes. They signed the same indemnity and waiver forms required of the local flight passengers.

Policemen on motorcycles cleared a path through the crowds for cars carrying the passengers to the mooring tower. The guests took the elevator to the landing platform, gingerly crossed the 2-foot gap to the swaying gangplank and headed down the dim, canvas enclosed corridor followed by stewards and luggage.

Before stepping into the elevator Wing Commander Colmore announced "We will be back next year as soon as possible after the snow clears" and termed the voyage home "a normal flight across the Atlantic." Shortly after 9 o'clock J.A. Wilson, Controller of Civil Aviation, Brigadier-General W.B.M. King and other officers and airport officials were in the mast head saying goodbye to the airship officers and crew members.

It would be a while before the Canadian passengers saw their homes again. For although the R100 intended to reach Cardington in 60 hours they would have to return by more conventional means. The number 13 was not superstitious to her officers as the airship was departing on August 13 with 13 passengers after 13 days in Canada. Adding those men making the round trip to Cardington (newspapers included Burney and Norway among the 13) there were 18 passengers.

At 9:30 p.m., while a large crowd gazed at the glow of lights from the ship's promenade deck windows and at the solitary figure on the gangplank waving goodbye, a bell sounded within the giant silver hull. Lights twinkled as the five motors started and the R100, held in the searchlight beams, slipped from the mast. The airship turned south and then passed over Montreal from west to east at such a low altitude that the passengers could read bread and brewery advertisements and even movie theatre names. The city "prickled with street lamps beneath the sign of the cross," observed C.H.J. Snider of Toronto's Evening Telegram, "was sliding under us like a velvet gem tray in a jeweller's window."

The sense of security the journalists felt from the ship's lack of motion was bolstered by the crew's nonchalance. No sooner had the R100 left the airport than off-duty crew members were playing jazz records bought in Montreal on the gramophone in their lower deck quarters.

During the airship's visit the Canadian National Railways carried 70,135 passengers from downtown Montreal to the airport, an estimated 527,908 arrived by car and approximately 40 planes provided joy rides, particularly during the two weekends, for 6,500 passengers.

Carrying 56 officers, crew and passengers (one watch was returning by sea), 9,585 gallons (32 tons) of fuel, 300 gallons of oil, 6.3 tons of water ballast and 1.5 tons of drinking water, the ship headed down the St. Lawrence River on three engines at 47 knots with a moderate 10-knot following wind. Depressions over the Atlantic meant a northerly crossing of the ocean to ensure a following wind.

Ships were moving slowly or were at anchor on the moonlit winding river but the R100 at her pressure height of 1,000 feet was not concerned with such physical limitations. It was a while before Montreal's great illuminated cross disappeared in the darkness. A newspaperman sheepishly claimed the first wireless message received: "Goodbye Gray Eyes — Joan." Passengers, crew members and their friends at the airport also took the opportunity to exchange farewell messages and good wishes for a successful flight.

Sorel was reached at 10:23 p.m. to the whistles of ships in the vicinity. About this time the newspapermen signed up for pumping duty. Pumping gasoline by hand from the main tanks up to the gravity tanks over the engine cars also helped to trim the ship. For the passengers it was an opportunity to get some relief from inactivity and to assist the crew; and the half-hour exercise, performed by two men on each 2-hour watch, would make excellent news copy.

The airship passed over Trois-Rivières at 10:50 p.m. and Quebec City at 11:45. There was a little motion in the ship in the vicinity of the Quebec capital but nothing serious. Under a clear sky and bright stars car headlights could be seen on the highways. At one small town on the river an attendant switched the lights of the local tennis court off and on to welcome the British visitor. The officer on watch in the control car with little to do flashed messages with an Aldis signalling lamp to ships and lighthouse keepers.

Jerry Snider dropped his radiogram in the wireless basket before retiring for the night. For the rest of the trip the newspapermen dropped their copy in a post bag four times a day to be collected by the wireless operator on duty at set times.

There were three or four dispatch times in each of the four wireless stages, depending on whether the ship was in touch with St. Hubert, Louisburg, Rugby or Cardington. If it was not possible to dispatch the messages in the time laid down they were to be returned to the sender and marked for priority by the officer of the watch for the next dispatch. The sender could alter the text but not exceed the number of words (60, 70, 80, 90 or 100) laid down for the next schedule.

The wireless stage in operation was indicated by a number placed over the press bag. The newspapermen were not allowed to talk to the wireless operator; any questions were to be referred to the officer of the watch.

The ship's officers had said that on an airship "you sleep like a log and eat like an elephant." No one had trouble falling asleep in his sleeping bag which held mattress, sheets, blanket and eiderdown. Father Point near Rimouski was passed at 2:05 a.m. As an experiment the Royal Canadian Corps of Signals' Ottawa station broadcast on shortwave for 10 minutes in every hour until the R100 was definitely out of range. The transmission faded out at 4:57 a.m.

By daylight the tail wind became light and, therefore, of little help to the airship. The western end of Anticosti Island was reached at 6:15 a.m. Three minutes before, with signals from both vessel and St. Hubert becoming fainter and static increasing steadily, the airport had teletyped Louisburg to take over the wireless traffic. The Nova Scotia station at 6:27 a.m. reported signals were very good.

At 6:40 a.m., 600 miles and 9 hours from St. Hubert, the airship lost touch with the airport and picked up Louisburg. While over the island the ship picked up the Rugby, England station.

Minor problems arose in the morning before some passengers tested their hardy appetites. Some of the taps above the washbasins worked poorly or not at all; the basins were made of duralumin which made them difficult to clean after use with soap; and there were no showers; a steward provided the hot water for shaving.

Breakfast of bacon, eggs, toast, marmalade, and coffee or tea was served in the dining saloon on tables nailed to the floor, the wicker chairs were movable. The attractive crockery was made of a celluloid material.

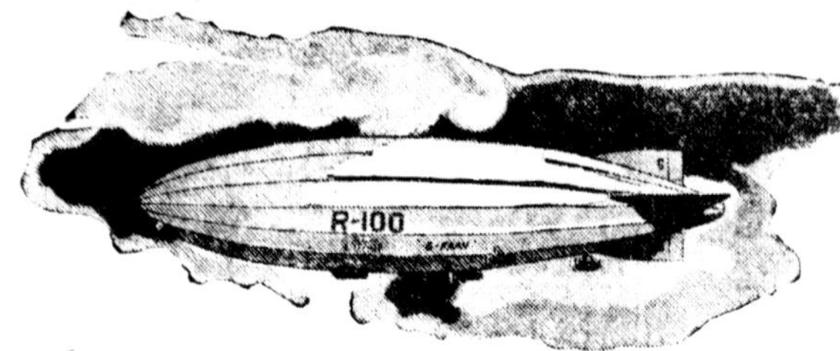

R-100 Food Expert . . .

. . . takes back supply of Chase & Sanborn's Fresh Coffee

THE following comment on Chase & Sanborn's Products from this food expert, aboard the R-100, is a tribute worthy of reproduction:

"Even before we arrived in Canada," says Chief Steward Savidge, "we had a taste of some of the good things in store for us . . . Chase & Sanborn's 'Fresh' Coffee, for instance. Throughout the voyage, everyone in the dining saloon of the R-100 enjoyed the genuine pleasure of drinking Canada's nationally preferred coffee.

"Being true Britishers, however, we had to have our tea as well, and Chase & Sanborn's Tea proved equally as delightful as their coffee.

"We are, of course, taking back a complete supply of each on the return voyage."

STANDARD BRANDS Daily Delivery Service . . . the largest fresh food delivery system in Canada . . . brings Chase & Sanborn's Tea and Coffee "fresh" to grocers, twice each week. On your table their exquisite, fresh flavor will bring the same enjoyment that has made them the choice of connoisseurs everywhere.

CHIEF STEWARD SAVIDGE of the R-100, for many years Steward on the "Majestic," served during the war with the British Submarines in Russian waters and on the North Sea. More recently, in England, it has been his privilege to serve many famous personages, among them H.R.H. The Prince of Wales and the King of Afghanistan.

Chase & Sanborn's TEA *and* COFFEE

Roasted in Canada

Although the R100 appeared in many advertisements this is the only product directly endorsed by one of her company.

Group Captain Stedman witnessed half-dressed passengers trooping through the dining saloon on the way to the washroom while others were trying to eat their breakfast. He thought this arrangement might be acceptable on an experimental airship but would never be permitted on a commercial line.

At 8 a.m. the ship informed the Air Ministry that she was headed for Belle Isle. Many vessels in the Gulf of St. Lawrence greeted the R100 as she headed for Quebec's Cape Whittle including, at 9 a.m., the Canadian Pacific ship *Montclare* which was almost directly below her. The airship was doing about 44 knots on three engines with a following wind of about 40 knots.

The weather was fine with good visibility and no low clouds. A report from Cardington forecast that winds would be westerly for the entire Atlantic crossing. The north shore of Newfoundland's Great Northern peninsula was a barren land of rocks, lakes, little vegetation and a couple of fishing hamlets. In the Strait of Belle Isle were several small icebergs and three Newfoundland Banks schooners.

At 11:54 a.m. the ship passed Belle Isle and headed east true at 60 knots ground speed with a light following wind; the sky was clear with some cirrus cloud visible and the ocean was like a sheet of glass. However, she was in the middle of a two-hour transmission silence. Louisburg reported by teletype to St. Hubert that the ship appeared to be in a "dead pocket", as she could neither hear other stations nor be heard herself.

The R100 had covered the 850 miles from Montreal in 13½ hours. The journey so far had been uneventful although one engine blew a spark plug as the Atlantic crossing began. To some passengers the upper surfaces of the numerous icebergs off the Newfoundland coast were like "polished white marble and the sea around was blue as sky." At 2:30 p.m., 142 miles east of Belle Isle, the R100 was sighted by the 22,000-ton Canadian Pacific liner *Empress of Australia*, two days out of Quebec City. Some newsmen were starting to get bored, not knowing what to do with themselves.

About an hour before sunset the airship, still cruising at about 45 knots, passed under a wedge-shaped pall of cloud with its point to the north. The following wind was light, around 10 knots, and the eastern sky very dull. A little later when at about 1,700 feet, the ship encountered some rain and collected water for ballast.

The ship altered course to the northeast in the hope of picking up a fair wind at a point in 55°N and 35°W. Soon the sun, shaped like "a loaf of bread, all molten gold", quickly slipped below the horizon leaving crimson clouds to reflect its light over the ocean in a long afterglow.

Cirrostratus cloud indicated a depression was approaching from the southwest. Soon the airship, now about 500 miles out on the ocean, encountered a strong northeast wind which reduced her ground speed despite starting a fourth engine. Her weather map plotted from information from Louisburg during the evening showed a low area south of the town but not its direction.

A bell summoned the passengers to dinner. Send-off presents of flowers, mostly sweet peas, decorated the dining saloon. After dinner of soup, fish, vegetables, sweets, biscuits, cheese, peaches and coffee Wing Commander Colmore, Major Scott and Squadron Leader Johnston remained seated around the blue-clothed tables. Scott discussed with Jerry Snider of Toronto's Evening Telegram the R101's forthcoming trip to India — a navigating task much more exacting than the R100's round trip Atlantic crossing because it involved greatly varying temperatures and meteorological conditions affected by land, water, mountains and desert. He did not say the Indian trip was impossible, but certainly was not looking forward to it with confidence.

These senior airship men played a dice-and-domino game called Up the River or expressed their views on a regular Atlantic airship service. Limitations of such a service included airships being able to land only at fixed points which required expensive mooring masts and other facilities, although mid-ocean floating concrete service stations, either moored or maintained in position by powerful tugs, were suggested; and a weekly round trip across the Atlantic, while permitting three days ashore, would have high initial costs.

Burney, whose proposals for an Atlantic airship service were, of course, debated, wanted to return to Canada as soon as the R100 was reconditioned and take Prime Minister Bennett to the forthcoming Imperial Conference. The other Airship Guarantee Company

PAC RE-19795-5

Over the Atlantic: Chief navigator Squadron Leader E.L. Johnston taking sights from the observation platform atop R100's bow.

representative on board, N.S. Norway, had a long talk that evening with E.W. Stedman, Chief Aeronautical Engineer of the Department of National Defence, on the technical aspects of operating airships in Canada some time in the future.

After Johnston had announced the ship's position as latitude 53°N and longitude 42°W someone in the dining saloon volunteered: "If this keeps up we'll be breakfasting in Piccadilly Saturday morning." Scott could not resist adding, with a chuckle as he patted a bit of the mahogany veneer pillars, "if this doesn't keep up we won't want any breakfast."

Just at that moment the lights went out. Every man in the dining saloon was wondering how many seconds were left before the airship met the Atlantic. There was no rolling or pitching; the hum of the engines continued; and pressure within their ears told the passengers they were not losing height but gaining. When the two minutes of darkness ended Chief Steward Savidge, who had been standing behind Snider's chair, claimed the engineers were just changing over one engine. In the meantime Booth, the ship's captain, had disappeared from the room.

Altering course had not helped the airship. Around 10 p.m. the wind changed to East North East and the rain increased. The R100 had run into the north side of the depression.

Eldridge, the captain's clerk, wearing his flying suit and fur-lined boots, led Snider down the stairway, past the chartroom and through the crew's quarters out to the 12 inch-wide walkway for pumping duty. There in the vast interior with the continuous creaking of girders electric lights broke the darkness beneath the creased and sagging gas bags. Rain water ran through the walkway and disappeared into the bottom of the hull. Eldridge told his partner to grab the quarter-inch lifeline that ran on one side of the walkway if the lights went out again.

Gone was the gentle purring of the engines one heard in the dining saloon as they now worked on the shiny lever of the fuel pump. Standing with the two midship power cars on either side and the third one 90 feet aft they could see through the ventilators in the ship's envelope the red exhaust flames of the Rolls-Royce engines. Occasionally Snider and

PAC PMR-73-574

On the way to Cardington, Britain's air attaché in Washington, Wing Commander L.J.E. Twistleton-Wykeham-Fiennes, and Group Captain E.W. Stedman in rubber-soled shoes on the upper promenade deck.

Eldridge turned a flashlight on a dial below the walkway to find out when 400 gallons had been pumped.

The officer of the watch suddenly appeared gesturing wildly. He rightly feared the consequences of fuel from the overflowing tank meeting the engine exhausts. The pumping team of Fergus Grant and Jacques-Narcisse Cartier also encountered problems. An engineer from one of the power cars scrambled up to inquire why they were pumping fuel into the ocean. Someone had evidently opened the wrong valve.

Snider and Eldridge made their way back to the control car entrance just as the lights went out again. The two coxswains manned the steering wheels undisturbed in the green glow from radium painted instruments before the control car's emergency lights took over.

The wind howled and rain beat against the windows from which the ridges and furrows of the white fog banks floating just below seemed as solid as the grooved and rounded tops of the icebergs spotted earlier in the day. Visibility was nil and the airship's ground speed was down to 35 knots. Throughout the night she made not much more than 40 knots against the brisk northeasterly wind. In difficult weather the officers ran over into double and triple watches.

Giblett was at the chart table drawing in red and green isobars to indicate high and low pressure areas. Earlier he had held Johnston, the chief navigator and also in charge of the wireless section, responsible for allowing press dispatches to crowd out a weather report Giblett claimed would have revealed the course of the depression. The next meteorological report he received came from England and showed the low had moved northeast across the ship's path. Because the airship did not carry dual wireless transmission equipment she could not send and receive messages at the same time.

Snider asked Giblett if the ballast tanks were full. "Lord, yes," replied the meteorologist, "we've caught enough rain to float the British navy. It's through the fabric of the envelope and into the cabins. It's short-circuited the lights and blown out the electric range and the radiators. Savidge will be distinctly annoyed when he tries to boil the eggs for breakfast, and Booth is sore already, because he hates shaving in cold water."

Most passengers had turned in for the night. Their sense of adventure was wearing off as they slept late the next morning; they awoke to find a very wet ship. Most of the rain water which had saturated the linen walls of the unoccupied passenger cabins and the kitchen came down the exhaust trunks, the observation platform on top of the ship, the automatic and manoeuvring valve flaps, and between gas bags 5 and 6; the remainder poured through the top cover which was porous where the dope had cracked along the seams.

Meals were different now. Instead of bacon and eggs, toast and coffee, the 56 passengers and crew were served sardines, bread and butter, and beer or lime juice. Milk was the favourite breakfast beverage; the beer soon gave out. A typical hot lunch of beefsteak and onions with two vegetables and peaches now became jellied chicken and salad, and canned fruit with lime juice, milk, lemonade or whiskey and soda. Chief Steward Savidge regretted having to serve canned beef and throwing spoiled lamb chops overboard. The lack of hot tea was regarded as a tragedy.

By daylight a moderate gale was blowing. In order to get the westerly winds the airship altered course to the southeast to get on the south side of the depression. Most of the passengers first learned of the storm and its effect on the electric range when they asked for shaving water. The weather had had no effect on the ship's motion. If passengers wanted physical evidence that the airship was changing altitude they placed a pencil on the floor and watched it roll.

Around 8:30 a.m. the rise in air temperature, from 48°F to 63°F, and very disturbed air for a short while indicated the ship had passed into the warm sector in the southern part of the depression. Confirmation came from First Officer Captain George Meager who, using a periscopic drift sight and a paper cube of aluminum dust dropped onto the heavy running sea, found the wind was due west at 40 knots. Half an hour later the clouds lifted and continuing to run on four engines she resumed her east true course at 82 knots ground speed.

PAC RE-20940-1

Dropping water ballast. The 180 foot long, 18 inch wide walkway linked the dining saloon and the bow exit. A 12 inch wide walkway aft of the passenger compartment ran for 130 feet to the two engine cars and 90 feet further to the third car before ending at the fins and rudders.

At 9:25 a.m. wireless communication was established on 2013 metres with Portishead 1,400 miles away on the Somerset coast. Unlike the outward voyage the R100 was in almost constant touch with land stations. Louisburg stayed with the ship until 33°W. and even exchanged signals with her at 26°W. Thus, for some time she was in touch with both sides of the Atlantic.

Just before noon Captain Meager climbed the 80 foot steel wire ladder to check the ship's latitude by shooting the sun. When his calculations indicated the airship was 60 miles north of her estimated position he believed his sightings with a bubble sextant were in error. But the navigator, Squadron Leader Johnston, who had seen him descending the ladder, checked Meager's figures and announced that they confirmed his own earlier sightings. Clouds obscuring the water prevented the navigator from taking drift measurements which would have indicated a change in wind direction; thus, the ship was considerably off course.

The day's sweepstake, organized by the captain's clerk, on the airship's run from noon to noon won N.S. Norway $6.75 in Canadian money for his guess of 1,250 nautical miles.

Water was still streaming into the very damp and warm ship. The R100, cruising at 52 knots on four engines, was still flying in thick cloud and occasionally descended to about 800 feet to determine the wind's direction (now south about 40 knots) by drift measurements. At 12:45 p.m. the weather cleared a little; a cloudy sky and the rough sea could be seen. The strong wind, almost on the airship's beam, produced a drift of about 30° to 40° giving her guests the impression the craft was driving sideways.

The newspapermen seemed to show little interest in the control car. Although Booth had ruled that only three of the passengers could be in the control car at one time seldom were there three.

The R100 had intended to head for the south of the Irish Free State but around 1 p.m. with the wind South South West the officers decided to lay a course for the north of Ireland. By 4 p.m. the weather had moderated considerably and after passing over dense stratus the

PAC RE- 13642

Group Captain Stedman monitored airship activity in Germany, England and the United States for the Department of National Defence. Figure 170 behind him is the course being steered.

ship sailed into a clear patch where sightings were taken. Around this time the wind veered to West South West so the officers decided to make landfall at the south of Ireland after all.

At 4:30 p.m. the airship, making good about 67 knots in bright clear weather, was about 600 nautical miles from landfall. She received wireless bearings at about latitude 52° 40′ N, longitude 23°W from the *Ausonia*, which had left Quebec City Friday, August 8, bound for Southampton. At 6:30 the R100 passed over Canadian Pacific's *Beaverbrae* the first vessel seen since leaving the Strait of Belle Isle.

The airship expected to reach Fastnet Rock, the most southwesterly point of Ireland, at 2 a.m. During the evening the ship was making 60 to 70 knots and except for some low cloud the weather was good. Her shadow was leading the way like a dark balloon.

At 1:25 a.m. Saturday, August 16 Bull Rock appeared on the port bow and 21 minutes later the Fastnet was sighted. At 2:12 in 50°N, 9°W the airship was in direct communication with Cardington, 365 miles away.

First Officer Captain Meager turned in at midnight. His quarters were now in the duty officer's cabin above the control car as his cabin on the return flight had been given to newspaperman Fergus Grant. He awoke several hours later with pressure in his ears which told him that the airship was changing altitude rapidly. Sensing something wrong Meager jumped out of his bunk still wearing his Teddy flying suit and entered the control car to find the officer on watch, Squadron Leader Booth, with his hands on the height steering wheel. The R100 captain was cursing that the rigger on wheel duty, E.R. Patterson, had allowed the airship to descend to 500 feet above the ocean. Soon the ship was once more at 1,500 feet.

Lundy Island in the brown waters of the Bristol Channel was reached at 7:45 (British Summer Time). A tan-sailed trading ketch with a white foresail was "glowing like a ruby in the rising sun" on the foam-crested water. The fifth engine was put on giving the airship about 65 knots including the 10-knot following wind. The welcoming roar from a large steamer's whistle a thousand feet below made shaving water quiver.

Ilfracombe appeared on the starboard side. The view of the distant north shore of Devon and Somerset was a welcome contrast to the previous day's water and clouds. The channel narrowed and the Welsh cities of Barry and Cardiff appeared on the port side.

The airship climbed to 2,700 feet. Coastal steamers in the Severn River estuary welcomed the R100 with their sirens. The English coast was crossed at Avonmouth. The factory whistles of nearby Bristol greeted the ship as she passed over at 9:15. Two airplanes from Filton aerodrome near the city provided an escort for a while. With breakfast over, the passengers were captivated by the English countryside with its irregular-shaped and hedged fields painted green, then yellow, cows which scattered at the sight of the plump shadow while sheep ignored the visitor, and cars which appeared as black ants on white ribbons under the bright Saturday morning sun and somewhat bumpy air.

The two huge Cardington sheds were spotted at 10:47 a.m. In one shed Britain's other experimental airship, the R101, was still in two pieces. Upon seeing Bedford Dennistoun Burney announced "About home" as he went through the motions of lighting an imaginary cigarette.

In the control car stood Squadron Leader Johnston wearing his blue uniform and service ribbons. On his left Flying Officer Steff pulled small cords which opened valves to release hydrogen from the gas bags. On his right Squadron Leader Wann controlled the five engines by signalling the engineers in the three power cars.

With the radiotelephone around his neck Johnston tried to reach the mast to ask if the station was prepared to moor the airship. Cardington ended his monologue with words of welcome and praise.

Johnston was compelled to interrupt and ask again: "Are you prepared to moor the ship? If we don't get down now we shall have to buzz around all day till the cool of the evening." The station was ready. The engines acknowledged commands through the control car dials.

The long quarter circle approach and descent towards the mooring mast gave one reporter his most noticeable example of motion on the entire return flight — his typewriter began crawling across the table! While the airship was manoeuvring into position to land, passengers heard several sharp noises. Major Scott later explained the sounds as the movement of the gas bags on the fabric ceiling of the passenger compartment.

Steff released the last hydrogen flap, thereby answering the question of whether the ship could moor in the middle of the day. Nine hundred feet of the main wire, released from the bow at 11:30 a.m., were allowed to trail on the ground in order to discharge static electricity accumulated during the flight. The landing party coupled the ship's wire to the one laid out on the ground from the top of the mooring mast, water ballast was released, and then the yaw guy ropes were lowered. As the mast was winding in the main wire the riggers lowered the ropes for attaching to the stern rollers so the crew could disembark quickly. The men of the R100 were just as anxious as the passengers to find cigarettes and hot tea. An air base official handed each man and passenger a package of cigarettes and matches as he emerged from the mast.

At 12:06 Acting Tower Landing Officer N.G. Atherstone reported "Secure." After 2,955 nautical miles and 57 hours, 36 minutes the R100 had completed her 10th flight, arriving home at a pressure height of 4,200 ft. and with 3,270 gallons (11 tons) of fuel still remaining. She consumed 8½ tons less fuel than on the flight to Canada.

During the R100's final hour of flight the number of cars near the Cardington mast had increased from about 50 to 200. While a large crowd at St. Hubert wishing the R100 bon voyage had pushed the total airport attendance during the Canadian visit past the 600,000 mark, only several hundred persons, mostly relatives and friends of the crew, welcomed her home. Among them were Lord Thomson, the Secretary of State for Air, and Sir W. Sefton Brancker, his Director of Civil Aviation. But it was not the great welcome N.S. Norway had been expecting. "We slink in unhonoured and unsung in the English style . . . ."

A watch composed of R101 crewmen and First Officer Atherstone went on board to relieve the R100 crew. While she was being refuelled supports under fuel tanks in the forward part of the ship gave way, allowing the tanks to fall into the bottom of the hull where they tore a hole in the outer cover.

A landing party assembled early the next morning to walk the ship into her shed. Work to be done included reconditioning the five engines and repairing the silent starboard forward engine, replacing the elevator fabric, part of which had failed over the St. Lawrence River, and inspecting the damage caused by the falling fuel tanks.

Once in her shed a closer examination revealed that the R100 could not fly again until her outer cover and gas bags were replaced — at a cost of £100,000. No decision had been rendered on whether the ship would have an additional bay, which could accommodate the 4 ton weight of a new cover and allow more fuel to be carried. Wing Commander Colmore later suggested that if the R100 could have flown at a constant 65 knots the voyage to Canada might have been reduced to 2½ days and the homeward journey to under two days. He considered dual wireless reception and transmission — to handle the large amount of weather, direction-finding and press traffic — essential navigation and meteorological equipment for an airship crossing the Atlantic.

The cost to the Department of National Defence of the R100 visit including her postponed May visit was $62,562.18. Material and labour for making hydrogen gas were $25,000, half that amount for the troops' pay and transportation and the balance for gasoline, radio and teletype services, floodlighting, the public address system, telephone charges, tent floors and horse stalls.

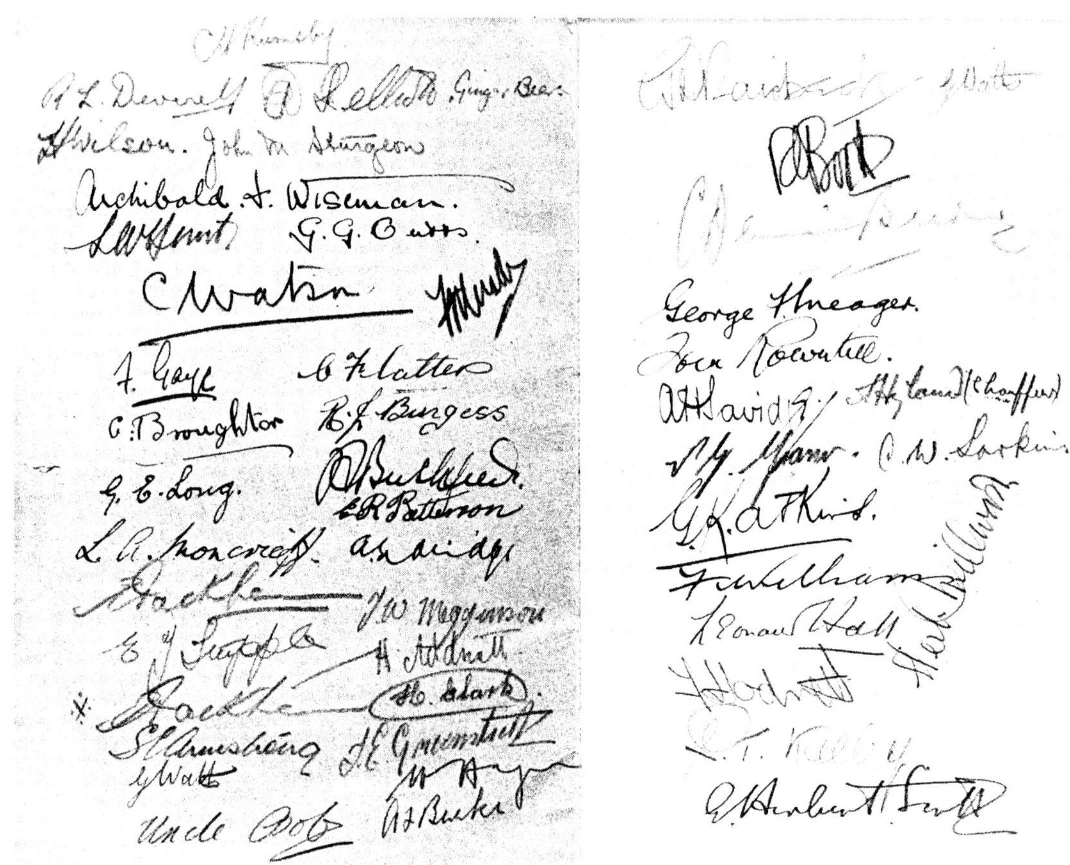

Signatures of R100 officers, crew and guests on paying-off dinner menu, Dujon Restaurant, Bedford, England, September 27, 1930. The Airship Guarantee Company, which had done less than £1,000 worth of business since the R100 launching, was about to suspend operations on November 30. N.S. Norway and chief designer Hessell Tiltman offered to bring the staff to Canada if some firm was interested in building or operating airships.

R101 in Cardington no. 1 shed on the left, R100 on the right. Probably September 30, 1930.

The End of British Airships

Airships, up to date, have neither proved a failure nor achieved an assured success.
Prime Minister Ramsay MacDonald
May 14, 1931

With the R100 safely home and undergoing repairs attention focussed on the R101 and her maiden flight to India.

The R101 incorporated many new design features, not all of them successfully. Her power cars and diesel engines, the only kind thought suitable for a flight to the tropics, weighed 17 tons. It was not possible to fit the five Beardmore 585 h.p. engines with variable pitch propellers to allow the engines to be reversed so four engines in separate power cars gave forward power.

She also had a steering system that functioned erratically. All these new devices added to the weight of the airship. Instead of having a disposable lift of 60 tons, as specified in her contract, the R101 had only 35 tons. The R100 had a disposable lift of 57 tons. Both airships, less power plant installations, weighed just over 95 tons although the original specifications had called for a fixed weight of 90 tons including power plants.

On November 28, 1929 the Air Ministry decided to give the R101 more lift by cutting the airship in half and adding another bay. It was also decided to enlarge the gas bag wiring to allow more gas to be pumped into each bag. The division took place on July 29, 1930. The R101 became the largest airship in the world, 777 feet long, on August 21.

The success of the R100's flight to Canada was a spur to Air Ministry officials. They were under pressure from the public, who were anxious to see if the "socialist" airship could do as well as the R100, and from the press, who had been critical of the many delays and setbacks.

Lord Thomson, Secretary of State for Air and an enthusiastic supporter of the airship program, was determined to travel on the R101 on her maiden flight to India. He planned to return to England — by airplane if necessary — in time to attend the meeting of the Imperial Conference's Committee on Civil Aviation October 20.

Cardington was a busy place as the R101 took her final series of test flights on October 2. She was scheduled to leave for India on October 4. There was no time to complete the speed trials before the ship departed with her two-day-old Certificate of Airworthiness.

The weather outlook for the flight was not promising. Rain and wind were forecast for the first 12 hours. The R101's trials had been conducted in mainly fair weather and only one flight made with the new bay.

As the R101 left the mast at 6:36 p.m. on October 4, she dropped 4 tons of her 9¼ tons of water ballast and slowly gained height. She circled the nearby town of Bedford, passed low over London and crossed the English coast at Hastings at 9:35 p.m. heading for Paris.

Argosy of Dreams

Space dwindles, time quickens. Motherland and Canada come closer yet together as England's mighty Air-liner proves the practicability of inter-Empire flight.

Dreams of centuries come true, and with millions of people here and abroad, we join in thanks and congratulations to the British Air Ministry and the gallant officers and men of the R-100 on the successful completion of her voyage.

IMPERIAL OIL LIMITED

Serving Canada for 50 years

6 Refineries 1900 Branches Thousands of Dealers

Home from Canada but relegated to the back pages of some London dailies by the fifth and final cricket Test match between England and Australia. The British Broadcasting Corporation described the R100 arrival from the mooring mast.

Two days after the sole trial flight with her extra bay the R101 left for India October 4, 1930, but without Canadians Major-General McNaughton and Arthur Steel whom the Prime Minister would not spare from the Imperial Conference delegation for the voyage. She had been kept as a stand-by in case the R100 could not fly to Canada.

Britain's 1924 experimental airship program ended six years later on a hillside in northern France, October 5, 1930. Dense mass of girders at edge of woods marks R101's passenger compartment.

On board were Lord Thomson, Sir Sefton Brancker, Director of Civil Aviation, and other official guests representing the Royal Airship Works. Major Scott, who had recently accompanied the R100 to Canada, was in overall command.

The distinguished passengers retired to the smoking-room after supper and turned in after sighting the French coast. It had been raining steadily since the airship left Cardington and now the winds were gusting between 40 and 50 miles per hour. At 2 a.m. the airship was 45 miles north of Paris over the town of Beauvais. Several minutes later the airship went into a long dive that lasted 30 seconds. She struggled to an even keel but went into another dive almost immediately. The R101 struck the crest of a hill and her 5.5 million cubic feet of hydrogen caught fire. Out of the 54 persons aboard only 6 survived: four engineers, a wireless operator and a representative from the Royal Airship Works.

Although the casualties were but four more than when the airship R38 broke in two over the Humber River in 1921, the crash of the R101 struck Britain as no disaster had since the sinking of the *Titanic* 18 years before. Messages of condolence from around the world were sent to King George, Prime Minister MacDonald and the Air Ministry. Canadians first learned of the disaster from bulletins on the radio and from church pulpits on that Sunday morning. The nation remembered those aboard the R101 who had recently been in Canada with the R100 and recalled Lord Thomson who had travelled from coast to coast two years before as a member of the Empire Parliamentary Association tour. The Department of National Defence ordered flags at all armouries throughout Canada flown at half-mast as a tribute to the officers and men who had lost their lives in the disaster. Flags on city halls and public schools were also at half-mast.

Britain paid her respects to the airmen after a warship returned the bodies home. The 48 victims lay in state in Westminster Hall. Relatives of the deceased, the Prince of Wales, the cabinet and members of Parliament, ambassadors and Imperial Conference delegates attended the memorial service in St. Paul's Cathedral on October 11. During the service the R101's pennant, which had survived the fire, was displayed on the high altar.

The R100 never flew again, despite her successful flight to Canada. The tragedy over Beauvais ended Britain's involvement with airships although the final decision was not taken until the following year. On September 11, 1931 Prime Minister Ramsay MacDonald announced in his House of Commons speech on reductions in government spending that the Cardington base and the overseas mooring masts would be placed on "care and maintenance." A secretariat to monitor airship activity in other countries was now the basis of the nation's airship policy — a plan rejected by the government only the previous May 14. The R100, Britain's only airship, was to be disposed of.

On December 2 the Under-Secretary of State for Air, Sir Philip Sassoon, told the House of Commons that preserving the R100's framework would have required approximately £1,000 a year and revealed that before deciding to scrap the airship, she had been offered for sale to an unnamed overseas government.

The demolition of the R100 began that same day. The duralumin frame of the airship, without gas bags or engines, was sold for £450 to the London firm of Elton, Levy and Company for scrap. The frame was attacked with axes. The workmen, some of whom had been crew members, had to cut the huge fins to length. Some of the longitudinal girders were just unbolted and allowed to drop to the floor many feet below. In the end a steam roller flattened these duralumin pieces for packing into bales ready to be shipped to London for smelting.

One bay between transverse frames 11 and 12 was retained for structural testing purposes and great care was taken with the passenger compartment whose tables and staircases appeared substantial but were, in fact, lessons in weight saving. Some of the bunks were sold to yachtsmen and to those with a sentimental attachment to the R100.

Canada maintained the St. Hubert mast for several years in the hope that other airships would use it. On November 3, 1937 an order-in-council authorized the dismantling and disposal of the St. Hubert mooring mast, now the property of the Department of Transport. The mast's winches went by tender competition to Manseau Shipyards Limited of Sorel, Quebec for $4,500 while M. Zagerman and Company Limited of Ottawa was awarded demolition of the structure for $4,875. The tower, judged a hazard to planes using St.

R100's bow was allowed to drop for easier dismantling in Cardington shed no. 1. She had taken over the R101's shed October 1, 1930 in preparation for inserting an additional bay; the workshops were adjacent. The Air Ministry had estimated the R100 reconditioning, including about 6 months for the manufacture of a new outer cover, would take until June 1931.

Demolition began at Cardington December 2, 1931. R100's control car and three-storey passenger and crew compartment above can be clearly seen. Canvas strips (aft of the compartment in this photograph) separated the gas bags to prevent them from surging in rough weather.

Hubert Airport, was dynamited at 1 p.m., January 13, 1938 and toppled into the snow. Backfilling of the hydrogen pipeline on June 15 left no reminder of the mooring mast which had cost $376,000 to build and was used only once.

Chopping R100's lower fin to pieces, December 3, 1931. On the right is a power car without its engines.

POSTSCRIPT

Three years of preparations which began with an offer to help Britain's experimental airship program ended with a two-week visit by an airship intended for a London-Egypt commercial air route. But Canada can recall her welcome to the British visitor in many ways. For the R100 remains in records, poems, buttons, postcards and even pieces of fabric. After the R100's dismantling one of her two steering wheels was presented to J.A. Wilson, Controller of Civil Aviation, and is now in the National Museum of Science and Technology, Ottawa.

Fourteen paintings depicting a history of transportation to 1931 remain in the former head office of the Canadian Imperial Bank of Commerce (then the Canadian Bank of Commerce) at 25 King Street West in Toronto. One showing the R100 approaching the very same building was taken down during renovations in 1972-73 and cannot be found.

A 15 foot model of the R100 built in 1969 for the Ontario Science Centre, Toronto, remains on display there.

Canada's first international airport is now Canadian Forces Base St. Hubert.

R100

She was alive — and they hated her,
Willed her as dead as their wits!
They knew how highly men rated her —
So now she is lying in bits.
For she was the work of their masters,
Whose mastery put them to shame,
She gave them no dower of disasters
So they made her one — hacksaw and flame!

Four hundred thousands of money,
Two generations of brain,
And the ways and the means and the money
Have all to be called for again!
Airworthy, tried and found trusty,
She showed up a ship that was not,
They knew she would never rot rusty —
So a steamroller rolls o'er the lot!

Now start from a newer beginning,
Apostles of lighter-than-air —
Such battles are well worth the winning
And ultimate vict'ry is there.
But when art and its opulent neighbour
Have triumphed with sore-blunted tools,
Do not trust the fine fruits of their labour
To the spite of poor panic struck fools.

J.H.

From Lancashire Aero Club Magazine, *The Elevator*

Postcards

R-100 Circling over Montreal, Canada.

Appendix A

VOYAGE TO CANADA
OFFICERS AND CREW OF R100

Captain	Squadron Leader R.S. Booth
1st Officer	Captain G.F. Meager
2nd Officer†	Flying Officer M.H. Steff
Navigator†	Squadron Leader E.L. Johnston
Meteorologist†	M.A. Giblett
Supernumerary Officer	Squadron Leader A.H. Wann
Chief Coxswain	Flight Sergeant T.E. Greenstreet
Chief Engineer	W.Y. Angus
Chief W/T Operator†	S.T. Keeley
Chief Steward†	A.H. Savidge
Captain's Clerk	A. Eldridge

Watches

No. 1

Riggers
Asst. Cox'n G.E. Long
C.H. Rumsby**
R.L. Deverell
G.G. Cutts**
C. Broughton

Engineers
Chargehand
E.J. Stupple**
H. Millward**
F. Gaye
H. Clark
J.M. Sturgeon**
R. Ball**

W/T Operators
A. Disley‡

Cook
J. Meegan

No. 2

Riggers
Asst. Cox'n T. Hobbs
S.C. Armstrong*
A.F. Wiseman
D.M. Kershaw*
E.R. Patterson*

Engineers
Chargehand
N. Mann
J. Jowitt
H. Cumley
L. Hunt
H. Addinell*
D. Lelliott**

W/T Operators
G.K. Atkins†

Steward
F. Hodnett†

No. 3

Riggers
Asst. Cox'n L.A. Moncrieff
R. Burgess*
F. Williams**
G.R. Scott
C. Flatters**

Engineers
Chargehand
G. Watts
D.L. Simmonds*
A.F. Burke*
C. Watson*
L. Hall*
H. Wilson*

W/T Operators
C.W. Larkins*

* relief watch in Canada, May-August 1930
** returned by sea, August 1930
† killed in R101 crash October 5, 1930 as were passengers Wing Commander R.B.B. Colmore and Major G.H. Scott
‡ survived R101 crash

Appendix B
Some Statistics of R100

Managing Director .... Sir Charles Dennistoun Burney
Chief Engineer and Chief Designer .... Barnes Neville Wallis
designer at Vickers Aviation in Weybridge, 1930
Assistant Chief Engineer and Designer .... J.E. Temple
Chief Calculator and Chief Engineer (1930) .... Nevil Shute Norway
Fabrics Department and Hydrogen Plant .... Philip L. Teed
Works Manager .... James Watson
Assistant Works Manager .... W.A. Dove
Engine Installation .... A.E. Palmer
Chief Draftsman .... William Horrocks
Chief Designer (1930) .... Hessell Tiltman

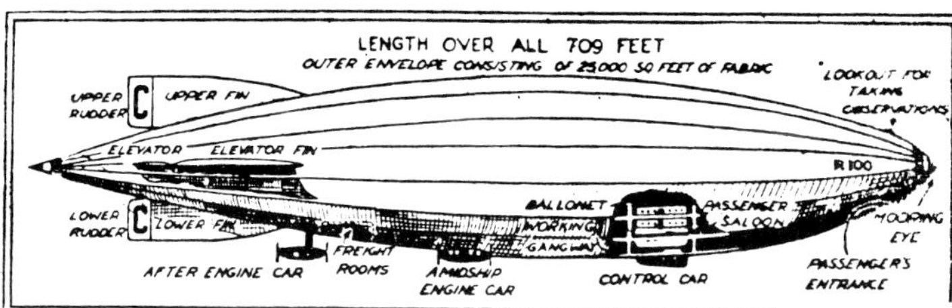

Constructed by the Airship Guarantee Company at Howden, East Yorkshire, England

Length: Dec. 1929 .... 709′ 9½″
June 1930 .... 695′ 5¾″
Diameter, maximum .... 133′ 4″
Height, with bumper bag .... 138′ 8″
Volume, hydrogen .... 5,156,000 cu.ft.
Gross lift .... 156.52 long tons
68 lbs of lift per 1,000 cu.ft., 97% gas purity
Fixed weight .... 105.52 long tons
Disposable lift .... 51 long tons
Speed, maximum .... 81 mph
Speed, cruising .... 65 mph
Ratio, length to diameter
Dec. 1929 .... 5.32:1
June 1930 .... 5.21:1
No. of gas bags 15
largest (no. 7): 551,890 cu.ft.; total area: 55,800 sq.yds.
Fin length .... 125 ft.
starboard, port, upper extend from frames 12 to 15; lower fin starts halfway between frames 12 & 13.
Total fin area .... 11,400 sq.ft.
each of above 3,100 sq.ft.; lower fin & rudder 2,100 sq.ft.
Engines 6
Rolls-Royce Condor IIIB 12 cyl., 650 b.h.p. RAF service in Avro 549 Aldershot III single-engine bombers, No. 99 Squadron at Bircham Newton 1924-26
Total weight of 3 power cars with 3 Bristol gas starters & accessories .... 10.189 tons
2 cars at frame 9, one at frame 12
Normal max. gasoline capacity ... 10,500 imp. gals.
(42 tanks, each 250 gals.)
Water ballast .... 18 tons
10 tons carried in 18 bags for quick discharge or max. 16 tons at a mast for slow discharge
Drinking and washing water .... 4 tons
(4 tanks, each 1 ton)
Accommodation .... 100 passengers
18 (4 berth), 14 (2 berth) cabins on 2 decks

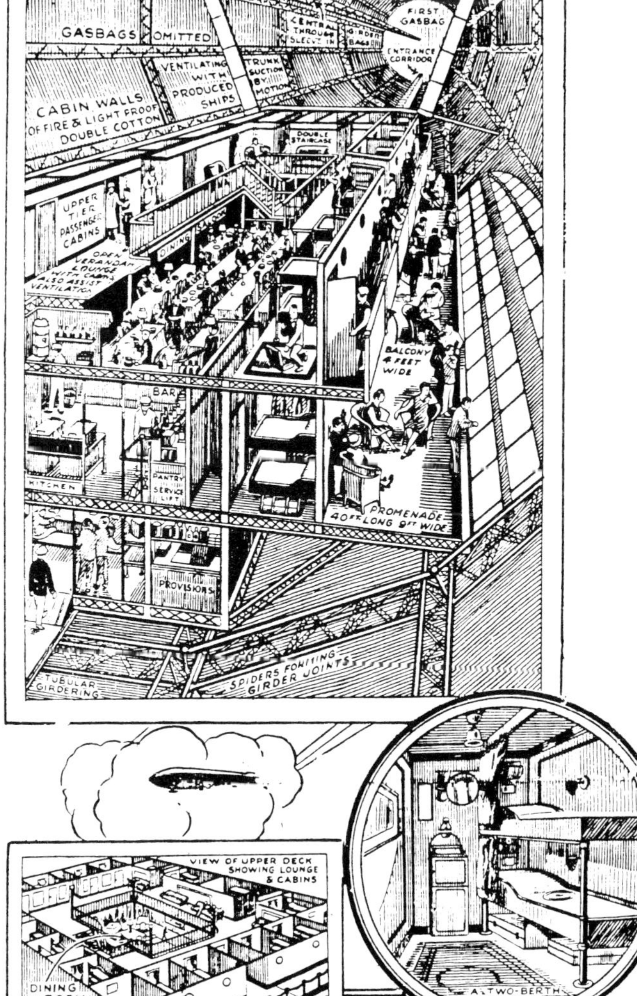

Appendix C
Suggested Signals Between Tower And Airship When Landing

On Ground	**Day**	**Night**
Main wire coupled	White flag waving	White flashing light
Main wire missed	Red flag	Red light
Starboard yaw guy coupled	Green flag	Green fixed light
Port yaw guy coupled	White flag	White fixed light
In Ship — Before coupling on		
Couple on if possible	White flag	White light
Don't attempt to couple on	Red Very light	Red Very light
In Ship — After coupling on		
Heave in as requisite	White flag	White fixed light
Stop heaving in	White flag withdrawn	White light flashing then out when acknowledged
Veer all wires	Red flag	Red light
Stop veering	Red flag withdrawn	Red light out
On Tower		
Heaving in	White flag	White fixed light
Stopped	White flag withdrawn	White light out
Veering	Red flag	Red light

Ground signals will be made from the landing T. for the main wire and from a position near the foot of the mast for the yaw guys.

Ship signals will be made out of the control car and should be repeated from the bow mooring position.

Tower signals will be made from some selected positions so that the signalling light cannot be mistaken for standing mast lights.

— from the London conference August 11, 1927 on the standardization of mooring towers attended by the Air Ministry, the Airship Guarantee Company, the Zeppelin Company and two U.S. military attachés as observers.

Appendix D
R100 Flights

TRIAL FLIGHT	FROM	TO	DEPARTURE DATE	DEPARTURE TIME	MOORED DATE	MOORED TIME	DURATION HRS-MIN	DISTANCE NAUTICAL MI.
1 Relocation	Howden	Cardington	16/12/29	0753	16/12/29	1340	5-47	130
2 Speed & Controls	Cardington	Cardington	17/12/29	0947	17/12/29	1616	6-29	174
3 Speed & Turning	"	"	16/1/30	0920	16/1/30	2240	13-36	217
4 Outer Cover	"	"	20/1/30	0910	20/1/30	1620	7-18	214
5 Endurance & W/T	"	"	27/1/30	0938	29/1/30	1630	53-52	1780
6 Outer Cover & Engines	"	"	21/5/30	1900	22/5/30	1750	22-50	691
7 Outer Cover & W/T	"	"	25/7/30	1902	26/7/30	1918	24-16	805
8 Overseas	"	Montreal	29/7/30	0248	1/8/30	0537	78-49	3364
9 Southern Ontario	Montreal	"	10/8/30	1818	11/8/30	2015	25-57	805
10 Homeward	"	Cardington	13/8/30	2130	16/8/30	1106	57-36	2955

Eastern Daylight Saving Time for Montreal; all others G.M.T.

R100 housed December 18, 1929-January 16, 1930; January 30-April 24; April 27-May 21; May 26-July 25; August 17-October 1, 1930, when transferred to R101 shed; deflated December 11, 1930.

Source: George Meager, *My Airship Flights 1915-1930* and Sir Peter Masefield.

Bibliography

The Public Archives of Canada, Ottawa, hold most of the primary R100 and airship documents I have consulted and the papers of R.B. Bennett, W.L. Mackenzie King, Ernest Lapointe, A.G.L. McNaughton, J.L. Ralston, W.A. Steel, E.A. Weir, J.A. Wilson, and the diaries of G.J. Desbarats and Mackenzie King.

Air Board Reports 1920-22, King's Printer, Ottawa.
Ollivier, Maurice (ed). The Colonial and Imperial Conferences from 1887 to 1937. Vols. II and III. Queen's Printer, Ottawa, 1954.
Departmental Reports (External Affairs, Marine, National Defence, Public Works) 1926-30, King's Printer, Ottawa.
Hansard (Canada and Great Britain).

The Evening Telegram (Toronto); Gazette (Montreal); Globe (Toronto); Halifax Herald; Hamilton Spectator; Mail and Empire (Toronto); Manitoba Free Press (Winnipeg); Montreal Daily Star; Ottawa Evening Citizen; Toronto Daily Star.
The Daily Mail; The Times (London).
The Detroit Free Press; New York Times.

Abbott, Patrick. Airship. The story of R34 and the first east-west crossing of the Atlantic by air. Charles Scribner's Sons, New York, 1973.
The Advertising Club of Montreal. Visit of His Majesty's Airship R-100. Canada — 1930; and 8-page supplement dated July 28, 1930.
(The Airship Guarantee Company.) Airship R.100. Howden, Yorkshire, c.1928; 23 pp.
Ashbolt, A.H. "An Imperial Airship Service." United Empire. July, 1921, pp. 498-502.
Boothby, F.L.M. "Airships for the Empire." United Empire, March, 1924, pp. 154-165.
Brancker, Sir Sefton. "Progress in Civil Aviation." The Empire Club of Canada, Toronto, March 22, 1926.
Brigham, R.B. "R100." Journal of the Royal Aeronautical Society, February, 1930, pp. 184-189.
Burney, Sir Charles Dennistoun. The World, the Air and the Future. Alfred A. Knopf, London, 1929.
"The British Imperial Airship Scheme." Aviation, March 5, 1923, pp. 266-268.
"England to U.S. by Airship." Aero Digest, March, 1928, pp. 358-360.
"Empire Air Communications." United Empire, March, 1930, pp. 135-142.
"Aviation and the Empire." The Canadian Club of Toronto, August 7, 1930.
"The Story of the R-100." Canadian Geographical Journal (now Canadian Geographic), October, 1930, pp. 524-552. Text reprinted January, 1973, pp. 24-35.
Cave-Browne-Cave, T.R. "Safety from Fire in Airships." The Royal Aeronautical Society, London, 1927. Aeronautical reprint no. 23.
"The Machinery Installation of Airship R.101." Journal of the Royal Aeronautical Society, March, 1929, pp. 175-206.
The Connecting File, October, 1930 — monthly magazine of the Royal Canadian Regiment.
"The St. Hubert Mooring Tower — A New Type of Construction Work." Contract Record and Engineering Review, Toronto, November 21, 1928, pp. 1207-11.
"Further Data on St. Hubert Mooring Tower." Contract Record and Engineering Review, Toronto, November 28, 1928, pp. 1233-34.
Corriveau, R. de B. "The St. Hubert Airship Mooring Tower." The Engineering Journal (Engineering Institute of Canada), April, 1930, pp. 277-281.
Elworthy, R.T. "Helium in Canada." Department of Mines. No. 679. King's Printer, Ottawa, 1926.
Ferguson, J.D. "Sky-High With R-100." The Blue Bell (Bell Telephone), September, 1930, pp. 1, 4-6.
Flight magazine.

The Goat, August, 1930 — monthly magazine of the Royal Canadian Dragoons.
Grant, J. Fergus. "R-100." Canadian Vickers, Spring, 1966, pp. 3-5.
Higham, Robin. The British Rigid Airship, 1908-1931. G.T. Foulis, London, 1961.
Hitchins, F.H. Air Board, Canadian Air Force and Royal Canadian Air Force. Canadian War Museum Paper No. 2, Ottawa, 1972.
"An Improved Airship Mooring System." Babcock and Wilcox, January, 1929. 12 pp. (Reprinted from The Engineer, December 28, 1928 with additions.)
Johnston, E.L. "The Atlantic Flight of 'R.100'." Aircraft Engineering, November, 1930, pp. 287-289.
Kipling, Rudyard. "With the Night Mail. A story of 2000 A.D." from Actions and Reactions. The Macmillan Company of Canada, Toronto, 1909, pp. 111-142.
Leasor, James. The Millionth Chance. The story of the R101. Hamish Hamilton, London, 1957.
Longstaff, Alan N. "Telling the World of R-100." Canadian National Railways Magazine, September 1930, pp. 14-15, 39.
Macmillan, Norman. Sir Sefton Brancker. William Heinemann, London, 1935.
Maitland, Edward M. The Log of H.M.A. R34: Journey to America and back. Hodder and Stoughton, London, (1920).
McLennan, J.C. "Report on Some Sources of Helium in the British Empire." Department of Mines. Bulletin no. 31. King's Printer, Ottawa, 1920.
Meager, George. My Airship Flights 1915-1930. William Kimber, London, 1970.
"The Montreal Branch and R 100." United Empire, September, 1930, pp. 502-504.
Morpurgo, J.E. Barnes Wallis. Longman, London, 1972.
Nansen, Fridtjof. "The proposed Arctic Expedition in the Graf Zeppelin." The Geographical Journal (The Royal Geographical Society), January, 1930, pp. 67-70.
"Neon" (Mrs. Marion Acworth). The Great Delusion. A study of aircraft in peace and war. Ernest Benn, London, 1927.
Pratt, H.B. Commercial Airships. Thomas Nelson and Sons, London, 1920.
Rhoades, Guy E. "The St. Hubert Airport." Canadian Air Review, March, 1929, pp. 15-16.
Richmond, V.C. "Organisation of a Colonial Airship Service." Journal of the Royal Aeronautical Society, November, 1921, pp. 588-615.
Roberts, Leslie. "Mooring for the R100." Maclean's Magazine, May 15, 1930, pp. 7, 43-44.
Robinson, Douglas H. Giants in the Sky. University of Washington Press, Seattle, 1973.
" 'R-100', latest British Airliner." Scientific American, April, 1928, pp. 334-335.
Scott, G.H. "Handling and Mooring of Airships." The Royal Aeronautical Society, London, 1929. Aeronautical reprint no. 49.
"The Shuttle of Empire — The Airship." Empire Club of Canada, Toronto, August 6, 1930.
Shute, Nevil. Slide Rule. William Heinemann, London, 1954.
(N.S. Norway) "R.100 Canadian Flight, 1930." Journal of the Royal Aeronautical Society, May, 1931, pp. 401-414.
Smith, Julian. Nevil Shute. Twayne Publishers, Boston, 1976. (Twayne's English Authors Series).
Snider, C.H.J. "Across in R-100." Queen's Quarterly (Queen's University, Kingston, Ontario), Summer, 1931, pp. 417-432.
Spanner, E.F. About Airships. E.F. Spanner, London, 1929.
This Airship Business. Williams and Norgate, London, 1927.
Gentlemen prefer Aeroplanes! E.F. Spanner, London, 1928.
Stedman, E.W. From Boxkite to Jet. Canadian War Museum Paper No. 1, Ottawa, 1972.
"The Return Voyage of R100 from Montreal to Cardington." Canadian Defence Quarterly, October, 1930, pp. 50-56.
"Rigid Airships." The Engineering Journal (Engineering Institute of Canada), February, 1930, pp. 104-124.
Sueter, Murray F. Airmen or Noahs. Sir Isaac Pitman and Sons, London, 1928.
Sutherland, Duke of. "Airships and the Empire." Empire Review, December, 1923, pp. 1351-56.

Swettenham, John. McNaughton. Volume I 1887-1939. The Ryerson Press, Toronto, 1968.
Taylor; John W.R. "The Trailblazer they steamrollered" (R100). Meccano Magazine, December, 1965, pp. 8-10, 12.
Teed, P.L. "The First Inflation of R100." Aircraft Engineering, June, 1930, pp. 135-136.
Templewood, Viscount (Sir Samuel Hoare). Empire of the Air. The Advent of the Air Age 1922-29. Collins, London, 1957.
Thomson, Lord. Air Facts and Problems. George H. Doran, New York, 1927.
"Visit of the R 100 to Canada." The Engineering Journal (Engineering Institute of Canada), September, 1930, pp. 562-564.
Weir, E. Austin. The Struggle for National Broadcasting in Canada. McClelland and Stewart, Toronto, 1965.
Williams, T.B. Airship Pilot No. 28. William Kimber, London, 1974.
Williams, Tom. "The Visit of the R-100". Canadian Flight, November-December, 1978, pp. 16-17.

PHOTOGRAPH CREDITS:
Archives Nationales du Québec 56
Bell Canada 60
Mrs. Booth 1A, 46
Canadian National Railways 25, 57, 64, 74, 97
Canadian Pacific Railway 12, 13, 14, 32, 36
City of Toronto Archives 88 (bottom)
Crown Copyright 40
David Cook 119
Flight International 18, 33, 37, 38, 116(top), 118
The Illustrated London News Picture Library 114, 116 (bottom)
Mrs. Meager via Major J.A. Booth 79
Metropolitan Toronto Library Board 46, 98, 99, 104, 115
National Museum of Science and Technology, Ottawa 65, 72, 88 (top), 94
Naval Historical Center, U.S. Navy 8
Mrs. Pressey 26
Mrs. C.H. Rumsby via Reg Wilkinson 45, 89 (top), 96
Toronto Telegram Photograph Collection, York University Archives, Toronto 41, 42, 55, 91, 92
Vickers 14, 16, 28, 30
Reg Wilkinson 112 (bottom)